Civil Engineering Heritage
Eastern and Central England

Edited by E. A. Labrum, ERD, CEng, FICE

Other books in the Civil Engineering Heritage series:
Northern England. M. F. Barbey
Wales and Western England. W. J. Sivewright
Southern England. R. A. Otter

Future titles in the series:
Scotland
Ireland
London

Eastern and Central England
Published for the Institution of Civil Engineers by Thomas Telford Ltd,
Thomas Telford House, 1 Heron Quay, London E14 4JD

First published 1994

A CIP record exists for this book

ISBN 07277 1970 X

Typeset in Palatino 10/11.5 using Ventura 4 at Thomas Telford Services Ltd
Printed in Great Britain by Galliard (Printers) Ltd, Great Yarmouth, Norfolk

Preface

In 1971 the Institution of Civil Engineers set up the Panel for Historical Engineering Works to compile a library of records of civil engineering works of interest throughout the British Isles. The works covered by these records have been selected for their technical interest, innovation, durability or visual attraction. These archives have become the principal national repository for records of civil engineering works and are now regarded as the leading authority on historical engineering works. They are widely consulted by heritage organizations, the business community and those involved in private research. The records may be consulted by the public for research purposes on application to the Librarian.

The series *Civil Engineering Heritage* seeks to make available to a wider public information contained in these archives and research undertaken by the Institution of Civil Engineers' Panel for Historical Engineering Works.

Britain has a heritage of civil engineering structures and works, unrivalled anywhere. The skills of past engineers are in evidence throughout the land in an enthralling variety of bridges, structures, services and lines of communication, the infrastructure of society.

It is hoped that the great range of works covered in this volume will widen the interest and understanding of the reader in the immense variety of expertise which features in our civil engineering heritage. In recent years, people have become more aware of the value of their heritage and their responsibility to keep the best examples for future generations. It is hoped that the book will also sharpen the resolve of those who seek to advance the cause of conservation.

The Editor would like to thank the following members of the Institution of Civil Engineers' Panel for Historical Engineering works, who have been the principal contributors to this volume: B. M. J. Barton, J. A. Carter, R. Cragg, P. S. M. Cross-Rudkin, P. Dunkerley, J. K. Gardiner, R. V. Hughes, J. B. Powell.

The Panel's Vice-Chairman, T. B. O'Loughlin, has provided a valuable service in checking the text and advising on the selection of illustrations. His wide experience in preparing books for the Heritage series has contributed in great measure to the preparation of this volume. The Librarian of the Institution of Civil Engineers, M. Chrimes, has also given valuable assistance in the preparation of the text.

Thanks are due to various local authorities, public utilities and others, too numerous to mention in detail, for their help, and to the Ordnance Survey for permission to reproduce the maps.

Ted Labrum

Contents

Series Production Editor and Designer: Sally M. Smith

Front cover: Welwyn Viaduct. Courtesy of Aerofilms

Title page: Bishop Bridge, Norwich, 13th century

Introduction

The area covered in this volume reaches from the Humber to the Thames and from East Anglia to central England, excluding the great industrial area around Birmingham, which is described in the volume on Wales and Western England. The area to the north is covered in the volume on Northern England and to the south in the volume on Southern England. The eastern counties are mainly flat, with little interruption from hills, and include the distinctive area of the Fens, where large stretches lie below sea-level. Drainage works and flood control feature as special tasks for engineers in this region. A series of large rivers flow eastwards across the country: the Trent, Witham, Welland, Nene and Great Ouse. Towards the western side of the area, the country is more abrupt, with the incursion of the Pennines at the northern edge and the Chilterns in the south. The region is crossed by many national lines of communication, main roads, such as the Great North Road and Watling Street, and rail routes to the North of England and to Scotland. Many canals mark the era of the early Industrial Revolution, with the growing need to transport freight between the surging industrial power of the Midlands and the capital. The establishment of these routes involved enormous works and civil engineers played leading roles in their construction.

The material in this volume has been collated from much more detailed reports prepared by members of the Institution of Civil Engineers' Panel for Historical Engineering Works in various parts of the area. The Panel has attempted to cover the prime examples of our engineering heritage, the best examples of different types of structure, those which illustrate a special degree of innovation or initiative and works which achieve a special aesthetic quality while still fulfilling the intended purpose of a service to the community.

Included in the works described can be seen the achievements of famous names: Brindley, Jessop and Outram for canals, the Stephensons in building the world's first trunk railway and the Cubitts for the Great Northern line. Fenland called for particular skills in drainage and flood protection, initiated by Vermuyden. These great engineers portrayed in

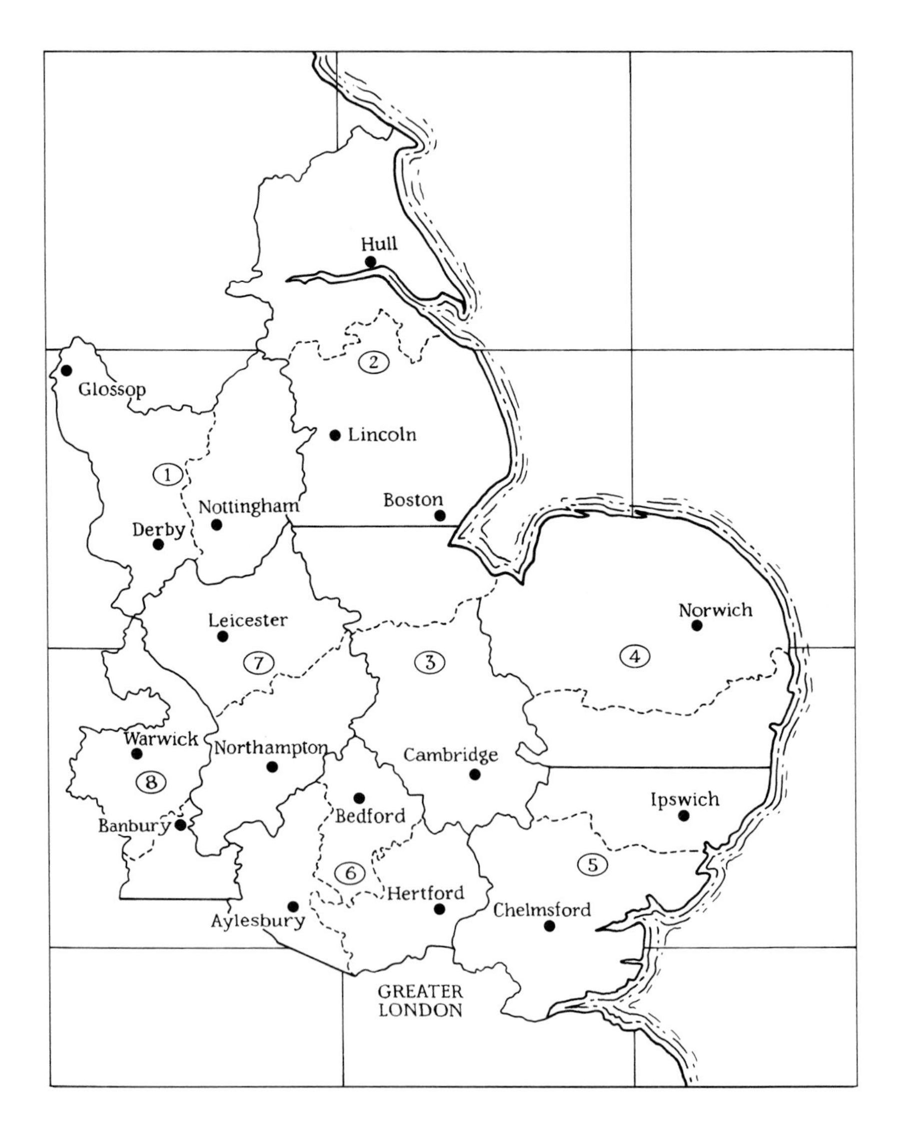
Hull
2
Glossop
Lincoln
1
Nottingham
Boston
Derby
Leicester
Norwich
7
3
4
Warwick
Northampton
Cambridge
8
Ipswich
Banbury
Bedford
5
6
Hertford
Chelmsford
Aylesbury
GREATER
LONDON

their works the spirit of enterprise of their times but there is plenty of evidence, in many other works described, of the skills and dedication of lesser known engineers and craftsmen.

This volume is divided into sections, broadly based on counties, except where it has been thought necessary to subdivide county areas. Each section is accompanied by a map of the region and a list of the works described together with a brief account of the background to civil engineering in the area.

The Historical Engineering Work (HEW) number of the Panel record, which is kept in the library of the Institution of Civil Engineers in London is quoted for each item.

An index of sites is included together with a list of other HEWs of interest which have not been included. The names of engineers, architects and contractors who were involved in the engineering works described are also listed. In addition to the specific references quoted after some of the items there is a bibliography of publications of more general interest, applicable to the region.

EASTERN AND CENTRAL ENGLAND

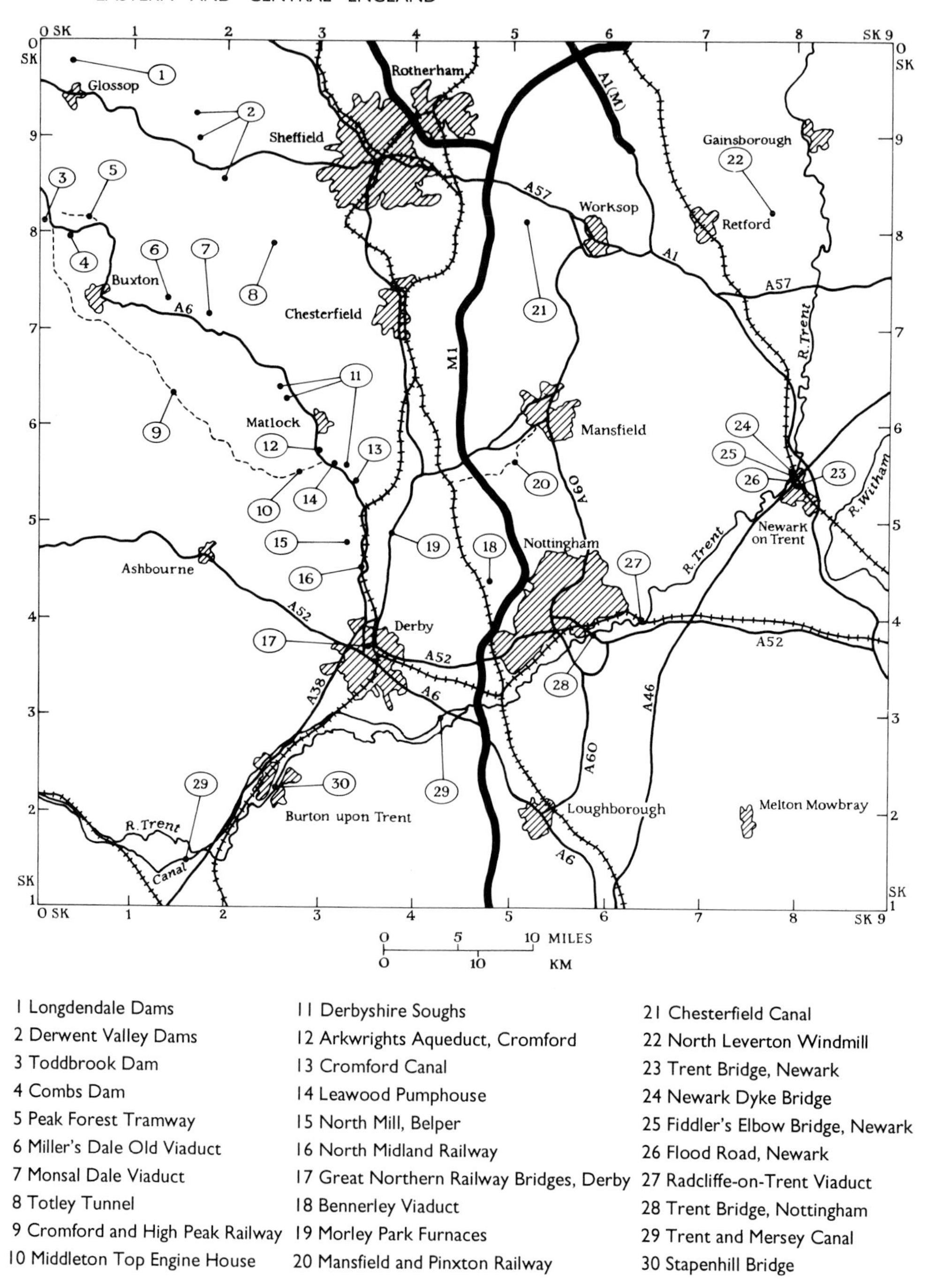

1 Longdendale Dams
2 Derwent Valley Dams
3 Toddbrook Dam
4 Combs Dam
5 Peak Forest Tramway
6 Miller's Dale Old Viaduct
7 Monsal Dale Viaduct
8 Totley Tunnel
9 Cromford and High Peak Railway
10 Middleton Top Engine House
11 Derbyshire Soughs
12 Arkwrights Aqueduct, Cromford
13 Cromford Canal
14 Leawood Pumphouse
15 North Mill, Belper
16 North Midland Railway
17 Great Northern Railway Bridges, Derby
18 Bennerley Viaduct
19 Morley Park Furnaces
20 Mansfield and Pinxton Railway
21 Chesterfield Canal
22 North Leverton Windmill
23 Trent Bridge, Newark
24 Newark Dyke Bridge
25 Fiddler's Elbow Bridge, Newark
26 Flood Road, Newark
27 Radcliffe-on-Trent Viaduct
28 Trent Bridge, Nottingham
29 Trent and Mersey Canal
30 Stapenhill Bridge

1. Derbyshire and Nottinghamshire

These two counties cover a great tract of country which changes gradually from the high hills and deep dales of the Peak District (the southern Pennines) on the west to the flat lands towards the lower Trent on the east. The oldest geological formation visible is the Carboniferous (mountain) limestone of 'The White Peak', which is overlaid successively eastwards by millstone grit ('The Dark Peak'), coal measures, Permian (magnesian) limestone, Bunter sandstone and Keuper marl, all of which strata dip to the east. Northwards the same kinds of country continues into Yorkshire, but southwards the hills descend into the Trent valley, which sweeps in a great curve eastwards and then northwards.

The Peak District is a home of the ancient industries of lead- mining (now defunct) and stone-quarrying (still proceeding), and, in the eighteenth century, textile manufacture was established in the Derwent valley. This territory has always presented difficulties to communications (roads, canals, railways) and has demanded particular skill on the part of civil engineers. The Pennines, with their deep valleys and high rainfall lend themselves to the making of impounding reservoirs for water supply and north-west Derbyshire has no less than 16 of them, named on the Ordnance Survey map. To the east, the Notts/Derbyshire coalfield, a continuation of the Yorkshire coalfield, runs from the Sheffield–Doncaster region in a broad band as far as the cities of Derby and Nottingham, and has supported a wide range of industries.

Still farther east and south a few colliery shafts penetrate the overlying strata to the 'concealed coalfield' beneath, but the land becomes more agricultural, like neighbouring Lincolnshire, although it also includes the Sherwood Forest area. The major industrial buildings are the power stations that stand at intervals along the banks of the River Trent.

1. Longdendale dams

HEW 1773

SK 00 97 to SK 10 99

Manchester water supply is taken by aqueduct and pipelines from a chain of five reservoirs 12 to 18 miles east of the city, which is fed by the River Etherow in Longdendale. An additional supply is drawn from a reservoir in a tributary valley. The whole project was designed by J. F. Bateman and provides a storage capacity of approximately 4500 million gallons. The work, executed during the period 1848–77, was carried out in two stages and represents a prime example of early Victorian puddle core dams. The upper three dams in Longdendale and two in a tributary valley formed stage 1.

Construction of the first dam at Woodhead began in 1848 on a site which had geological problems in the form of the Lawrence Edge landslide on the south slope of the valley and faulted rock, heavily fissured, on the north side. Below the river-bed Bateman founded the dam on a 16 ft deep cut-off trench which was filled with clay puddle and extended into the sides of the valley. This 10 ft wide puddle was carried up as the core of the 92 ft high embankment. The bank consisted of clayey material on the upstream side of the puddle and rock based material downstream. The dam was completed in 1851 but leakage developed below the puddle; this was reduced by an early form of grout injection but it later became necessary, in stage 2 of the scheme, to build a new embankment.

The site of the next dam, Torside, constructed between 1849 and 1854, is 2 miles downstream of Woodhead and was founded on more stable rock. The dam has a maximum height of 101 ft with a crest 890 ft long. The design of the embankment was similar to that of Woodhead with a central puddle core. The cut-off trench, 26 ft deep, turns upstream at the sides of the valley for about 200 yd before cutting into the hillsides, a favourite design feature of Bateman's practice.

The third dam, Rhodes Wood, was constructed in the period 1849–55 while work was already in progress on the other two dams. This dam stands 75 ft above the flood plain and was built to the same design as Torside. On the north side of the reservoir a flood water bypass channel, 36 ft wide and 10 ft deep, was constructed from Woodhead to Bottoms. This channel had to cross the toe of the

Tintwistle Knarr landslide and the slip was reactivated by the work. A rock buttress was built to support the slip while the channel sides were strengthened and the channel covered with an arch. Further problems developed in 1852 when heavy rains reactivated the Didsbury Intake landslide above Rhodes Wood Dam, causing damage to the spillway. The slope was successfully stabilised by drainage adits, a treatment approved by R. Stephenson and I. K. Brunel who were jointly consulted on the problem.

In the tributary valley, dams were built at Arnfield and Hollingworth between 1849 and 1854. Arnfield, which has a maximum height of 63 ft and a crest length of 3115 ft, still remains but Hollingworth was demolished in 1987.

The second stage of the scheme commenced in 1865 with construction of two further dams in Longdendale. These were Vale House, with a maximum crest height of 60 ft and Bottoms at 72 ft high.

The design of the dams followed that of the earlier ones but incorporated some improvements in the puddle cores. The cut-off trench for Vale House was taken down to a depth of 35 ft and the base of the trench sealed with concrete. An innovation in dam construction was the introduction of steam cranes for Bottoms Dam, which was finished in 1872.

The project was completed with the building of a second dam at Woodhead to eliminate the leakage problems of the earlier embankment. The new cut-off trench had to be taken down to a depth of about 100 ft on the south side to get below the toe of the Lawrence Edge landslide and to the even greater depth of 150 ft on the north side. The entire trench was filled with concrete, the first example of this solution in dam construction. On 7 February 1877 Bateman was able to report the succesful conclusion of this great project, after nearly 30 years of work.

Further reading

1. BATEMAN J. F. *History and description of the Manchester waterworks.* Manchester and London, 1884.

2. BINNIE G. M. *Early Victorian water engineers.* Thomas Telford Ltd, London, 1981, 171–189.

2. Derwent Valley dams: Howden Dam, Derwent Dam and Ladybower Dam

Howden Dam:
HEW 185
SK 170 924

Derwent Dam:
HEW 539
SK 173 898

Ladybower Dam:
HEW 540
SK 199 854

The upper end of the Derwent valley, in wild Pennine country north of all habitation, is occupied by three great reservoirs made by the Derwent Valley Water Board to supply Sheffield, Derby, Nottingham and Leicester.

The first two were made under the Derwent Valley Water Act of 1899 and the first to be completed was Howden, the most northerly one. Its masonry dam, with castellated towers near each end, has a length of 1080 ft, a height of 117 ft above the valley floor and thicknesses of 178 ft at the base tapering to 10 ft at the crest. Its overflow level is 870 ft OD. The reservoir has a capacity of 1980 million gallons, and the length from the dam to the head is 1¼ miles. Construction took place from 1901 to 1912.

A mile and a half downstream is the Derwent Dam, of similar design and dimensions, constructed from 1902 to 1916: length 1110 ft, height 114 ft, thicknesses as before, overflow level 776 ft OD. The reservoir has a capacity of 2120 million gallons and stretches to the foot of the Howden Dam.

The engineer for both these reservoirs was Edward Sandeman, who was Engineer to the Board. The construction was carried out by direct labour.

Between 1921–1930 a 1000-yd long tunnel was made under the ridge between the Derwent Reservoir and the Ashop valley to the west, and weirs were built in the valley, so that water from the River Ashop could be diverted to the reservoir.

A further Act of Parliament in 1920 empowered the Board to make the Ladybower Reservoir with a dam 3 miles down the valley from the Derwent Dam. The Ladybower Dam consists of an earth embankment with a clay core, under which a trench was excavated to an average depth of 180 ft (maximum 255 ft) and filled with concrete. The length of the embankment is 1250 ft, its maximum height 135 ft, and its greatest thickness 665 ft, tapering to 17 ft at the top. The overflow level is 668 ft OD. The capacity of the reservoir is 6300 mg; one arm of the reservoir stretches

Howden Dam, downstream face

nearly to the Derwent Dam and another arm stretches 2 miles up the Ashop valley. G. H. Hill & Sons were the engineers and the contractor was Richard Baillie of Haddington, East Lothian. Construction started in 1935 and was completed in 1943. Two viaducts were constructed to carry roads across arms of the reservoir and two villages (Ashopton and Derwent) were submerged.

The catchment area of all three reservoirs was originally 31 200 acres, but a further 7300 acres was added when the waters of the River Noe in Edale were diverted through another tunnel into the Ladybower reservoir in 1951.

Further reading

Derwent Valley Water Board brochure published on the opening of the Ladybower reservoir in 1945.

3. Toddbrook Dam

HEW 612

SK 007 811

Toddbrook Dam near Whaley Bridge was constructed to impound feed water for the Peak Forest Canal. Toddbrook Reservoir, which was brought into service in 1838, is supplied from a catchment area some 43 acres in extent. The reservoir has a total capacity of 300 million gallons and its longest stretch of open water is nearly 1 mile.

The dam is of earth construction with a puddle clay core,

having a top level of 614 ft OD. A new concrete spillway was constructed over the centre of the dam in 1971.

During routine operations in 1975, the water level in the reservoir was drawn down revealing a depression in the upstream pitched face of the dam. A routine inspection during 1977 revealed that the depression was settling at an annual rate of 5 inches. When the water level was drawn down further still, a second depression became visible. Following this discovery the reservoir was emptied in October 1977 for investigation, but extensive tests did not reveal the exact cause of the depression, so a clay blanket was added on the upstream bed of the reservoir to prevent further leakage and new rip-rap was placed. A wave-wall was added along the crest in 1981 and in 1983 extensive curtain grouting was carried out along the line of the core of the dam to seal an area at its base suspected of being porous.

The dam was refilled in 1984.

4. Combs Dam

HEW 609

SK 035 797

Combs Dam was designed by Benjamin Outram and was completed circa 1805. It was constructed by the Peak Forest Canal Company to supply the first feeder to their canal.

Combs Dam retains a reservoir of a notional capacity of 340 million gallons, supplied by five main streams flowing through a catchment area some 3094 acres in extent, which includes Combs Moss. The dam is an earth embankment 1000 ft long and it does not have a special clay core. The embankment reaches a maximum height of 52 ft above ground level.

The narrow crest of the dam was not level and a masonry wave wall some 1 ft to 4 ft in height supported the upstream face of the crest. The upstream face of the dam is protected by stone rip-rap; the vegetated downstream face has a slope of 1 in 1.6. The dam has a broad-crested overflow weir 241 ft long, a swallow hole spillway 4 ft 6 in. in diameter and low-level draw-off is by way of a 1 ft diameter cast-iron pipe which discharges into a 22 in. diameter culvert.

During recent years a new precast concrete wave-wall was constructed to give the dam a level crest and to allow it to cater for the possible maximum flood. A new berm was added on the downstream side.

5. Peak Forest Tramway

HEW 1683

SK 023 820 to SK 083 782

The Peak Forest Tramway extended some 6 miles from the Bugsworth Basin of the Peak Forest Canal to limestone quarries in the Dove Holes area. The Tramway was designed by Benjamin Outram and was built by Outram & Company between 1794–1797. The horse tramway was originally mostly single track, though it included a double track self-acting incline midway operated by rope and pulley. The downward loaded wagons were given a start on the steeper gradient at the top of the parabolic incline in order to haul up the empty wagons. About 1805 the line was doubled, except through the Chapel Milton Tunnel.

From 1794 to 1870, the track consisted of cast-iron plates 3 ft long supported on stone 'pot' sleeper blocks, with upstanding angles set to 4 ft 2 in. gauge. In 1870, wrought iron angles 9 ft long were substituted.

Nothing now remains of the track, but much of the route is traceable, and the Bugsworth Basin and quays on the canal are under course of restoration.

Further reading

SCHOFIELD R. B. Benjamin Outram. (1764–1805), canal engineer extraordinary. *Proc. Instn Civ. Engrs*, Part 1, 1979, **66**, Nov., 539–555.

6. Miller's Dale Old Viaduct

HEW 1383

SK 139 732

Built in 1862–63, this viaduct carried a double line of the Midland Railway 76 ft above the River Wye, the B6049 Buxton to Tideswell road and an occupation road in steep and heavily wooded country. It has masonry substructures and three openings of 87 ft 8 in. skew span (50 ft 6 in. square span) each consisting of five wrought iron arch ribs of 11 ft rise. Instead of skewbacks, the ribs thrust against horizontal plate web girders lying in cast-iron bedplates. In 1906, following the diversion of trains to a new steel truss bridge built alongside, the old viaduct was strengthened with steel bracing and a new deck. When the strengthening had been completed, traffic on the 'slow line' was diverted back over the old viaduct.

The railway was closed in 1968 but the old viaduct remains to carry a footpath (the Monsal Trail). A local preservation society, Peak Rail, hopes to reinstate a single line of railway across it.

7. Monsal Dale Viaduct

HEW 225

SK 183 715

A five-arch viaduct spans the narrow valley of the River Wye at Monsal Head, where the river abruptly changes direction from south-east to west. It carried the Midland Railway's Derby to Buxton and Manchester line, of which this length, Rowsley to Buxton, was opened in 1863. The railway emerged from Headstone Tunnel, three miles beyond Bakewell, immediately crossed this 300 ft long viaduct and headed north-westwards up the valley.

The river passes through the middle arch. The viaduct stands 80 ft above the river, but more than 100 ft below the overlooking hillsides, and the view looking down on it from the Monsal Head Hotel on the B6465 Bakewell to Tideswell road is justly famous, even though the viaduct is surpassed in size by many others.

Monsal Dale Viaduct (north face)

The viaduct was built in stone, except for the voussoirs on the face of each arch, which are in blue brick, but extensive patching has been done in red brick, with the result that a close view of the structure is less satisfying than a distant one. This route, which continues into Miller's Dale and Chee Dale was one of the most scenically spectacular railway routes in the country and, since the line was closed in 1968, it has become a walkers' route known as the Monsal Trail.

Although the Monsal Dale Viaduct is now generally accepted as part of the landscape, its construction (and indeed that of the whole railway) roused the wrath of the writer John Ruskin, who wrote a famous diatribe which included the words:

> 'There was a rocky valley, between Buxton and Bakewell, once upon a time, divine as the Vale of Tempe; you might have seen the Gods there morning and evening... You enterprised a railroad through the valley... The valley is gone and the Gods with it; and now every fool in Buxton can be at Bakewell in half an hour, and every fool in Bakewell at Buxton, which you think a lucrative process of exchange — you fools everywhere.'

The engineer of the viaduct was the Midland Railway's engineer, W. H. Barlow.

Further reading

COOPER B. *Transformation of a valley*. Heinemann, London, 1983 . 216–8.

LELEUX R. A Regional history of the railways of Great Britain, **9**, *The East Midlands*. David & Charles, Newton Abbot, 1976. 182–6.

8. Totley Tunnel

HEW 671

SK 251 788 to SK 306 802

Totley Tunnel, which is 3 miles 950 yd long is the longest railway tunnel under land (excluding the London Underground) in Great Britain; only the Severn Tunnel is longer. It was constructed from 1888 to 1893 on the Midland Railway's Dore and Chinley line which gave a more direct route between Manchester and Sheffield. This was the last trans-Pennine line to be constructed.

The west portal has an imposing horse-shoe shaped masonry arch carrying the inscription '1893 Totley Tunnel' and the M. R. monogram. It is at the foot of the steep western slope of the millstone-grit ridge between Millstone Edge and Froggatt Edge and can be viewed from the approach road to Grindleford station which crosses over the line within a few yards of the portal. The east portal at Totley is in a deep cutting and is not easily accessible.

Five shafts, up to 500 ft deep, were sunk in the construction of the tunnel. The work was troubled with vast quantities of water; a newspaper report said that 'every man seemed to possess the miraculous power of Moses, for whenever a rock was struck water sprang out of it'.

The engineer for the work was the firm of Parry & Storey of Nottingham and Derby and the contractor was Thomas Oliver of Horsham.

Further reading

RICKARD P. Tunnels on the Dore and Chinley Railway. *Min. Proc. Instn Civ. Eng.*, **cxvi**, 115.

9. Cromford and High Peak Railway

HEW 780

SK 314 558 to SK 012 816

The Cromford and High Peak Railway was constructed between 1826 and 1831 to connect the Cromford Canal, near its terminus at Cromford, to the Peak Forest Canal at its Whaley Bridge terminus, after an earlier proposal to connect them by another canal had come to nothing. Most of the 33-mile route is still easily traceable; a short part on the highest section of the line is still in use by British Rail as a goods line serving the Dowlow limeworks south of Buxton, although the route here has been extensively realigned.

The engineer of the line was Josias Jessop and the line was designed on canal principles — long level (or nearly level) stretches, connected by short steep inclines operated by cable haulage. In all there are eight levels, starting at 278 ft OD at the east end, rising to a summit of 1254 ft (on the fifth level), then falling to 517 ft at the west end. All the winding engines, and indeed all the ironwork of the railway, were supplied by the Butterley Company of Ripley, Derbyshire, of which William Jessop and Benjamin Outram (engineers of the Cromford and Peak Forest Canals respectively) had been partners.

Construction had barely started in 1826 when Josias Jessop died suddenly, but the work continued to his design under the supervision of his resident engineer and the railway was opened partially in 1830 and fully in 1831. Steam locomotives were in use from 1833 and pas-

Cromford and High Peak Railway, stone-faced embankment near Minninglow

sengers were carried from 1833 to 1876. It was a standard-gauge line and connections were made to other railways in due course. Most of the northern part of the line closed in 1892; the remainder closed in stages between 1952 and 1967.

Eighteen miles of the route from the eastern end is preserved as a public walkway and cycleway, known as the High Peak Trail. This length includes the wharf alongside the Cromford Canal (HEW 1778) at High Peak Junction where a transit goods shed, warehouse and workshop have been restored. The one remaining engine house and its machinery, also restored, is at Middleton Top (HEW 117). The line has two short tunnels and two cast-iron under bridges at Rise End (HEW 560, SK 278 552) and Longcliffe (HEW 287, SK 225 557), both 1865 replacements of the originals. Several of the embankments are stone-faced, one near Minninglow (SK 197 582) being 300 yd long and over 30 ft high in the centre. A commemorative stone bearing the name of Josias Jessop is to be seen on the north portal of the Newhaven Tunnel where the line passed under the Ashbourne to Buxton road (SK 151 630).

Another 3-mile length north of the summit level, partly running alongside Fernilee Reservoir, is also walk-

Middleton Top Engine House, looking west from the top of the incline

able and at the Whaley Bridge terminus stands a transit shed which the line entered at one end and the Peak Forest Canal still enters at the other.

Further reading

MARSHALL J. *The Cromford and High Peak Railway*. David & Charles, Newton Abbot, 1982.

RIMMER A. *The Cromford and High Peak Railway*. Oakwood Press, Blandford Forum, 1976. (Locomotion Papers, No. 10).

10. Middleton Top Engine House

HEW 117

SK 276 552

The only surviving winding-engines of the Cromford and High Peak Railway (HEW 780) have been restored to working order by the Derbyshire Archaeological Society and stand in the engine house at the top of the Middleton incline, where the line rose from 743 to 996 ft OD at a gradient of 1 in 8¼. The rail track has been lifted and only some large pulley wheels remain at the top and bottom of the incline as a reminder of the winding mechanism.

The engines are a pair of low-pressure condensing beam engines, originally steam-powered at 5 p.s.i., with a common crankshaft and flywheel. They were constructed by the Butterley Company in 1829.

The engine house has a semi-octagonal end, reminiscent of a toll house, looking down the incline. The boiler house, covered by a lean-to roof, is at the west end of the engine house and a tall chimney standing beside it forms a prominent landmark on the hill top. The two Cornish boilers are not the original ones and not all parts of the engine are original either, but the beams, parallel motions and connecting rods almost certainly are. The engine is run occasionally, using compressed air, as the boilers have not been reconditioned.

The whole assembly is scheduled as an Ancient Monument. It is in the care of Derbyshire County Council and is open to the public.

Further reading

MARSHALL J. *The Cromford and High Peak railway*. David and Charles, Newton Abbot, 1982, 18, 24, 26, 54, 58.

11. Derbyshire Soughs: Hillcarr Sough, Yatestoop Sough and Meerbrook Sough

Hillcarr Sough:
HEW 66
SK 257 637

Yatestoop Sough:
HEW 855
SK 264 626

Meerbrook Sough:
HEW 1721
SK 326 553

Sough (pronounced suff) is a local word for an adit driven nearly horizontally from a valley floor for the purpose of draining distant mine shafts. It has a gradient towards its outlet, usually of 10 to 20 ft/mile.

Thousands of lead mines were sunk in the Derbyshire limestone from Roman times to the 18th century, by which time the dewatering of the shafts had become a major problem and expense. Horse gins, water-wheels, hydraulic engines and steam engines were all in turn used for pumping, but the driving of a sough could greatly reduce the necessary pumping or even eliminate it. The longer soughs were often driven not by the mine-owners but by entrepreneurs who would take for payment a proportion of the value of all the ore that the sough had made accessible. The soughs were usually self-supporting, but where support was needed it was commonly provided by floors and roofs of stone slabs and sides of stone walling, often dry laid. All the lead mines have long been closed, but most of the soughs still drain water from the abandoned shafts. There were over 200 soughs, most of which are now lost and forgotten. The earliest known sough was driven near Cromford (not the present Cromford Sough) in 1629–36 by Sir Cornelius Vermuyden. Three soughs are especially noteworthy.

The longest sough was the Hillcarr Sough, of which the tail (outlet) is in Darley Dale, 300 yd west of the River Derwent. It is marked by a gritstone arch 8 ft high and 6 ft wide from which a copious flow of water runs to the river (SK 257 637). It was driven mainly between 1766 and 1787, and extended later, over a total length of 6100 yd, with many branches which probably doubled that length; it drained mine shafts from as far away as the Mosstone (or Mawstone) Mine south of Youlgreave.

Three-quarters of a mile to the south is the Yatestoop Sough, with its tail (an arch 5 ft high by 3 ft wide) a few yards from the river bank (SK 264 626). This is an older sough, started in 1743, which had been driven 2500 yd by 1764. It was extended in the 19th century to reach the mines at Elton, giving it a length of about 5000 yd. One of the mines served by this sough was the Mill Close Mine near Darley Bridge, which was once the richest lead mine in the country and continued working until 1940.

Six miles farther down the Derwent Valley, between Cromford and Whatstandwell, is the Meerbrook Sough, which has the grandest portal of all at its tail (SK 326 553) — an ashlar arch 10 ft wide and 7 ft high. It has a keystone inscribed FH 1772, the initials of Francis Hurt who owned lead mines at Wirksworth and was the principal proprietor of the sough. The date marked the start of the sough. The sough did not reach the terminus at Bolehill, less than 2 miles away, until 1845 — progress was interrupted for long periods of economic recession. The completion of the sough drastically reduced the flow in the Cromford sough (2 miles north), which had been used as a source of power by Arkwright's mills in Cromford and as a source of water by the Cromford Canal (HEW 1778). This interference resulted in the Arkwright Company having to abandon the two mills that depended on the sough and in the Canal Company having to build the Leawood Pumphouse (HEW 781) to augment their water supply. When all the lead mines served by the Meerbrook Sough had closed, it was bought by the Ilkeston & Heanor Water Board in 1903 and used for a public water supply, a use which still continues. Meerbrook Waterworks, recently modernised, stands near the sough tail, taking more than 10 million gallons a day.

Further reading

KIRKHAM N. Yatestoop Sough. *Peak District Mines Historical Society Bulletin*, 7, 1962, Oct. 5–21

FORD T. D. and RIEUWERTS J. H. (ed.). *Lead mining in the Peak District*. Peak District Mines Historical Society and Peak Park Planning Board, Bakewell, 1983. 96–97, 109, 119–121.

12. Arkwright's Aqueduct, Cromford

HEW 1474

SK 298 569

The warehouses of Sir Richard Arkwright's Upper Mill (1772) still stand fronting Mill Lane in Cromford. The road in front of the buildings is spanned by a small aqueduct constructed of ¾ in. cast-iron plates, having a total length of approximately 75 ft, supported on stone pillars. The longest span over the road is 22 ft with 14 ft headroom below. Its internal width is 3 ft 9 in. and height 2 ft 9 in., with occasional tie-bars across the top. The vertical joints between the plates are covered by iron pilasters 1 ft 4 in. wide, designed as an architectural feature. The continuous vertical plates forming the main span have a raised arch-and-keystone pattern cast in their external faces. The keystones bear the date 1821, when this structure replaced a timber one of 1785.

Most of the water needed to drive Arkwright's water-wheels in his cotton spinning mill was obtained from the Cromford Sough, one of many adits draining the shafts of distant lead mines. A high-level leat from the sough's outlet brought a flow of water (about 16 000 gallons per minute) to this aqueduct, which delivered it to an overshot wheel in the mill.

The completion of the Meerbrook Sough (HEW 1721), 2 miles down the valley, in 1845 so seriously diminished the flow in the Cromford Sough that it could no longer be utilised, and the aqueduct fell into disuse, but it still remains.

Further reading

COOPER B. *Transformation of a valley*. Heinemann, London, 1983. 69–70.

13. Cromford Canal

HEW 1778

SK 454 472 to SK 300 570

This was the first major canal (as distinct from river improvements) designed and supervised by William Jessop, assisted by Benjamin Outram. Completed in 1794, it ran 15 miles through Derbyshire from the head of the Erewash Canal at Langley Mill to Cromford, where Richard Arkwright had established his cotton mills, with a 2-mile branch to Pinxton. It had 14 locks, a long tunnel (2966 yd) under the Butterley ridge between the valleys of the Erewash and the Amber, a reservoir over the tunnel and two large aqueducts over the Rivers Amber and Derwent. The Butterley reservoir has a 300-yd earth dam (HEW 1615, SK 398 517) incorporating an early use of a clay-puddle core; it also has an outlet valve that discharges directly into the tunnel below. The locks on the canal, all below the level of the tunnel, were broad, but the tunnel itself and the canal beyond it were narrow.

The tunnel became impassable because of subsidence in 1900 and was never repaired, although traffic continued on the now detached part of the canal above the tunnel until the canal was formally abandoned in 1944. The uppermost mile of the canal from Cromford to the Derwent Aqueduct has been made navigable again and the restoration is being extended southwards. The east end of the Butterley tunnel is still visible, but the west end has been buried by roadworks and the bottom 2 miles of

Derwent Aqueduct (looking east)

the canal have been filled in and lost, despite the complete restoration of the Erewash canal from the Trent.

The aqueduct over the Amber has gone, but the Derwent Aqueduct (HEW 1484, SK 316 556) (also known as the Wigwell Aqueduct) still stands, a fine memorial to the canal's engineer. It carries the canal across the river on a segmental arch of 80 ft span, with an 18 ft rise from the springing; the top of the parapet walls are 38 ft above normal river level. There are two smaller arches, one on each river bank, and the total length of the structure is 400 ft. The aqueduct faces above the arch are relieved by two string courses. During construction in 1793 a crack developed in the arch; Jessop had it closed with tie-bars and the walls rebuilt at his own expense.

At the north end of the aqueduct stands the Leawood Pumphouse (HEW 781), built in 1849. In the same year the Ambergate to Rowsley railway was built and had to pass under the canal. At a point in the Leahurst estate an iron-trough aqueduct (HEW 284, SK 320 556) was inserted, the tensile stresses being taken by flat raking bars, which are visible on the external faces of the trough. The aqueduct is supported on masonry abutments and has a timber towpath with an iron balustrade alongside it. It is accessible by the canal towpath, about ½ mile south of

Leawood Pumphouse, Cromford Canal

the Derwent Aqueduct. The canal has been restored by the Cromford Canal Society from its terminus to the Derwent aqueduct, a distance of about 1½ miles.

Further reading

COOPER B. *Transformation of a valley*. Heinemann, London, 1983, 192–4.

HADFIELD C. and SKEMPTON A. W. *William Jessop, Engineer*. David and Charles, Newton Abbot, 1979, 40–44.

14. Leawood Pumphouse

HEW 781

SK 315 557

This is a gritstone structure of elegant design, housing a Watt-type beam engine, built in 1849 by the Cromford Canal Company for the purpose of raising water from the River Derwent to the long summit (and terminal) pound of the canal. It stands to a height of 45 ft on the right bank of the river, at the end of the Derwent Aqueduct, and has a 95 ft chimney stack with a large cast-iron cap.

The massive engine was designed and erected by Graham and Company of Milton Works, Elsecar, near Sheffield. The beam length is 33 ft; the piston has a diameter of 50 in. and a stroke of nearly 10 ft, working at 7 strokes per minute. The boiler house adjoins the engine house; the boilers are replacements, installed in 1900, with a pressure of 40 p.s.i. Water is drawn from the river through a 150 yd tunnel to a reservoir in the basement, whence it is lifted 30 ft and discharged into the canal.

The engine worked continuously until the canal was closed in 1944. It was restored, together with the boilers, by the Cromford Canal Society in 1979. It is now run periodically under steam and still lifts water to the canal.

Further reading

COOPER B. *Transformation of a valley*. Heinemann, London, 1983. 193–4.

15. North Mill, Belper

HEW 118

SK 346 481

Charles Bage built the world's first multi-storey building with an interior iron frame in 1796–97 at Castle Foregate Mill, Shrewsbury (HEW 425). William Strutt, the Derbyshire cotton magnate, was in touch with Bage and subsequently built himself a six-storey mill at Belper in 1803–4 on the foundations of an earlier mill destroyed by fire on the east bank of the River Derwent.

Like Bage's mill it has cast-iron beams and columns with the floors carried by brick arches spanning between the beams to reduce the risk of disastrous fire which all too often consumed the traditional timber-framed mills.

The North Mill, as it is called, is 123 ft long and 31 ft wide and still stands in reasonable condition, although it is now unoccupied. Also still in existence is a substantial masonry dam, which was built across the river to form a great mill pond from which an 18 ft water-wheel, once in the basement of the mill, took its power. The mill is overshadowed physically by the adjoining East Mill (later, larger and less elegant), but the North Mill is a fine example of industrial buildings in that period.

Further reading

JOHNSON H. R. and SKEMPTON A. W. William Strutt's Cotton Mills, 1793–1812. *Trans. Newcomen Soc.* 1956, **XXX**, 179–205.

SIVERWRIGHT W. J. *Civil engineering heritage Wales and Western England.* Thomas Telford, London, 1986,172.

16. North Midland Railway

HEW 1583

SK 33 to SE 33

This railway, which obtained its Act of Parliament in 1836 and was opened in 1840, was the first rail connection to Yorkshire from the south. It ran from Derby (which already had connections to Rugby) to Leeds, and its engineer was George Stephenson , who considered it to be the finest piece of railway engineering that he had executed. A 5-mile branch line was opened by a separate company from Rotherham (on the North Midland Railway (NMR)) to Sheffield. (The direct line from Chesterfield to Sheffield via Bradway tunnel was not opened until 1870.) The Manchester and Leeds Railway and the York and North Midland Railway both joined the NMR at Normanton, near Wakefield, approaching from the south-west and north-east respectively, making Normanton an important junction with a station disproportionately large for the size of the local community.

There were originally seven tunnels on the line, but three of them have since been opened out. Two of the remaining ones are of some interest. Toad Moor Tunnel at Ambergate (HEW 169, SK 349 514), constructed in gritstone blocks, is unusual in having an elliptical section with its major axis horizontal to resist the sideward pressure from a slippery, inclined bed of wet shale. Later some of the superincumbent mass was removed and the tunnel was lined with iron ribs. The tunnel portals are embellished with string courses of a roll pattern, which tended to be Stephenson's trade mark on this railway. Further north, Clay Cross Tunnel (HEW 121, SK 397 642), 1 mile 24 yd long, is in brick, but has an ornate castellated north portal in sandstone. The driving of this tunnel revealed rich coal seams which led Stephenson to establish coal mines and an ironworks there, from which the Clay Cross Company later developed.

Through the middle of Belper runs an impressive cutting lined with masonry retaining walls (HEW 1625, SK 348 477), up to 26 ft high, in two lengths of about 250 yd each, with closely spaced overbridges to match (there are five in one 270 yd length). There are cuttings over 50 ft deep in rock at Chevet, south of Wakefield and at Warmfield near Normanton.

Near Derby Station the NMR built an engine shed (HEW 1293) which is now used as a general workshop. It was known as the Round Shed but in fact is a 16-sided polygon, 140 ft across, with timber roof trusses supported on iron columns.

Further reading

ALLEN R. North Midland Railway guide. Nottingham, 1842, reprinted by Turntable Enterprises, Leeds, 1973.

WILLIAMS F. S. *The Midland Railway*, London, 1876.

North Midland Railway Belper Arches

17. Great Northern Railway Bridges, Derby: Derwent Bridge, Derby and Friar Gate Bridge

Derwent Bridge : HEW 170 SK 352 371

Friar Gate Bridge: HEW 119 SK 346 364

The Great Northern Railway (GNR)'s 'Derbyshire Extension' from Colwick, east of Nottingham, to Dove Junction, north of Burton-on-Trent, was opened in 1878 and closed in 1968. The engineer for the line was Richard Johnson, Chief Engineer of the GNR. His two notable bridges in Derby were erected by the Derby firm of Andrew Handyside & Co. to carry the line over the River Derwent and over the street called Friar Gate. The Derwent bridge has two wrought iron arch trusses of 140 ft span, crossing the river on the skew, with a rise at the centre of 16 ft. Each truss has four I-section girders (10 x 9 in., with webs horizontal), made of riveted curved plates and connected by cross bracing to form a composite truss of 2 ft 6 in. overall width and 5 ft depth. The deck at a level of 11 ft below the top of the arches is partly suspended from the arches in the central part and partly supported on struts off them towards each end. When the railway was in use a footway was cantilevered off one side of the bridge but now the whole bridge has become a footbridge and the cantilevered footway has been removed.

Top: Friar Gate Bridge, looking east, Creat Northern Railway

Friar Gate Bridge is in fact two bridges, side by side

but not quite parallel, which used to carry two double tracks into Friargate Station immediately to the west. They are highly decorated cast-iron structures, Listed Grade II and owned by Derby Corporation. The iron-work is painted in two tones of colour to emphasise the motifs of the decoration, with a brighter colour on focal points.

Each bridge has four segmental arch ribs of 72 ft span and 5 ft rise, each in five castings, with a depth of 3ft 6in. at the abutments and 2ft 6in. at the crown. The spandrels and parapets of the outside arch of each bridge have intricate cast patterns and so have the two outward-facing parapets. The soffits and transverse struts between the ribs, all visible from below, are also decorated. Projecting timber sidewalks above the spandrels rest on cast-iron brackets which also carry the parapets each of which comprises ten panels. The abutments are in plain sandstone.

Further reading

Engineering, 11 April 1879.

18. Bennerley Viaduct

HEW 120

SK 471 437 to SK 474 439

Bennerley Viaduct was one of several wrought iron railway viaducts built in the short period when this material had largely superseded cast iron and before it was in turn superseded by steel. Now Bennerley is one of the only two remaining viaducts of this type (the other is at Meldon in Devon, HEW 270).

It strides for 1/4 mile across the flat foor of the Erewash valley on the Notts–Derbyshire border, having 16 Warren girder spans of 77 ft mounted on tubular piers. The piers each comprise a group of 10 vertical wrought iron tubes, made up of quadrants with continuous longitudinal riveted flanges, with an additional raking tube at each side and with wrought-iron bolted cross-bracing, standing on concrete bases capped with bricks and gritstone. The piers support four lines of Warren girders, 8 ft deep. The decking is of corrugated troughs, which halved the quantity of ballast needed, at an almost constant height of 56 ft above the ground, surmounted by lattice parapets 26 ft apart. At the west end of the Warren spans are three

Morley Park Furnaces (1780 furnace right, 1818 furnace left)

iron girder spans, on brick piers, carrying the track across British Rail's Midland main line.

The viaduct was opened in 1878 by the Great Northern Railway as part of their Derbyshire extension, on the route from Nottingham Victoria to Derby Friargate. The GNR Chief Engineer at the time was Richard Johnson and the contractors Eastwood Swingler & Co. of Derby. The line was closed in 1965 and the approach embankments at each end have been removed.

Further reading

The Engineer, Oct. 19 1877.

19. Morley Park Furnaces

HEW 65

SK 380 492

Two square, truncated pyramids, built in gritstone, stand in a Derbyshire field, fenced round and silent, but in their day they were forceful symbols of the growing Industrial Revolution.

They were cold-blast coke-fired furnaces for iron smelting. The first was built in 1780, not long after the first-ever such furnace was built (1770) by Abraham Darby at Coalbrookdale, Shropshire, to which these furnaces bear a marked likeness. The second furnace was added in 1818. They are both about 40 ft high. The earlier furnace has its back to a natural cliff face and can no longer be entered; the later one stands clear of the cliff; inside it is like a bottle-shaped kiln and the top opening (for the charging of materials) is clear.

The first proprietor of the furnaces was Francis Hurt of Alderwasley (in the Derwent valley), who had inherited a family lead merchanting business and had branched into iron making. The furnaces last worked in 1874.

Access to the furnaces is from a minor road and via the drive to Iron Works Farm (passing under the A38), and thence southwards by a public footpath, from which they can be seen. They stand on private property.

20. Mansfield and Pinxton Railway

HEW 1775

SK 452 543 to SK 538 608

At the height of the canal age, Mansfield was without a canal connection because of its situation on high ground. A horse-tramway, with cast-iron edge rails, was laid over a distance of 8 miles from the town to a branch of William Jessop's Cromford Canal at Pinxton and was opened in 1819. The engineer was Josias Jessop.

In 1847 the Mansfield and Pinxton Railway (M & PR) was taken over by the Midland Railway (MR). The original track had been laid to a substandard gauge and the new owners set about converting this to the standard MR gauge, so that locomotives could operate along the line. The radii of the sharper curves was also eased.

Diversions of the route were made in 1871, 1892 and 1972 to improve the alignment. The line is still in use as a BR goods line, 5 miles of it being still on the original alignment. It is thus the oldest railway route in the East Midlands and is one of the oldest lines anywhere that has been in continuous commercial use since its opening.

A notable structure still standing is the King's Mill Viaduct (SK 520 598), 1 mile from the Mansfield terminus; it has five 24 ft stone arches and bears a date-stone 1817. The line no longer crosses it. One original crossing house still stands (SK 513 589) and an underbridge at Kirkby Woodhouse (SK 491 547) still retains its original appearance on its south side. The only remaining evidence of the canal wharf at Pinxton is the name 'Wharf Road' and the Boat Inn, still open but now far from any navigable water.

Further reading

BIRKS J. A. and COXON P. The Mansfield and Pinxton Railway. *Railway Magazine*, July, Aug. 1949.

21. Chesterfield Canal

HEW 1831

SK 387 717 to SK 786 946

The Chesterfield Canal is the oldest canal in the East Midlands (apart from the Trent and Mersey which was contemporary). It connected Chesterfield with the River Trent at West Stockwith, passing through Worksop and Retford. The first engineer was James Brindley and after his death in 1772 he was succeeded by his brother-in-law Hugh Henshall. Construction started in 1771 and the canal was opened in parts, being completed in 1777. It had 65 locks, one major tunnel, Norwood west of Worksop, and a short tunnel, Drakeholes east of Retford. The principal sources of water were the River Rother at Chesterfield, a series of reservoirs near Harthill for the summit pound, two other reservoirs further west, and several small rivers and streams.

Norwood tunnel is 2880 yd long and it was the longest in the country when it was completed in May 1775. A

Drakeholes Tunnel South Portal, Chesterfield Canal

section of the tunnel collapsed in 1908 but was not repaired and all traffic on the canal ceased in the 1950s. A canal society was formed in the 1970s, and the canal has been restored for navigation by pleasure craft between the Trent and Worksop. Between Worksop and Norwood, the canal still holds water but the locks are derelict; west of Norwood some of the line has been obliterated.

The canal caught the headlines one day in 1978 when a dredger working near Retford accidentally pulled up a wooden plug from a forgotten washout pipe and a long section of the canal quickly ran dry.

Further reading

HADFIELD C. *Canals of the East Midlands*. David and Charles, Newton Abbot, 1966.

22. North Leverton Windmill

HEW 770

SK 775 820

This is a sturdy three-storeyed four-sailed tower mill standing in the plain of the lower Trent. The tower is 38 ft high coated with tar and surmounted by a wooden ogee-shaped cap resting on a metal kerb and automatically rotated by a fantail to keep the sails turned to the wind. The patent sails are 62 ft from tip to tip and sweep close to the ground so the tower has no external gallery.

On the first floor are three pairs of millstones, two of Derbyshire gritstone for grinding grain for cattle and one of French quartz for grinding flour. A sail-powered hoist on the second floor is used for lifting sacks.

The mill was built with two storeys in 1813 and was heightened in 1884 to provide a third storey when a new cap, sails and machinery were installed. Periodic replacement of sails has gone on ever since. A diesel engine, for use when the wind fails, was installed about 1956. The mill is still working.

It was built as a subscription mill by a company formed by a group of local farmers. That company continued until 1956 when responsibility passed to a new company which, unusually, still operated the mill as a commercial undertaking. At the same time a voluntary organisation was formed for fund-raising. The mill is open to the public.

The original builder is not known. Thorntons of Retford,

North Leverton Windmill

Notts. carried out the 1884 heightening and R. Thompson & Son of Alford, Lincs carried out extensive repair work in 1961.

Further reading

BROWN R. J. *Windmills of England*. Robert Hale, London, 1976, 160–161.

23. Trent Bridge, Newark

HEW 1894

SK 796 541

The old Great North Road at Newark crosses the Newark Branch of the River Trent immediately downstream of Newark on a seven-arch stone bridge, sometimes called Newark Bridge. It was built in 1775 by Stephen Wright and widened in 1848 by adding cantilevered footways and iron railings on each side, supported by heavy, but decorative, iron brackets. There are pilasters on the piers between the arches and stone cutwaters below. On each side at centre span is a decorative lamp standard with a panel below bearing a coat of arms and the date MDCCCXLVIII.

The width of the bridge is 38 ft. The river is 170 ft wide which is considerably less than at Trent Bridge, Nottingham (23 miles upstream) because this section of the river only takes part of the flow. The major part goes along the Kelham branch which passes about 2 miles from the town.

24. Newark Dyke Bridge: Warren Truss and Whipple Murphy Truss

Warren Truss: HEW 1023

Whipple Murphy Truss: HEW 777

SK 801 558

Newark Dyke means the Newark navigation branch of the River Trent, which bifurcates upstream of the town and unites again downstream of it. The main line built by the Great Northern Railway crosses the left branch of the Trent, the Kelham branch, on a low brick-arch viaduct but crosses the Newark branch, the navigation channel, with a single span.

The original bridge comprised two separate pin-jointed Warren-girder structures one for the up line and one for the down line, with a span of 259 ft and an overall depth of 17 ft, described as the largest of their type at the time. The upper booms of cast iron were hollow and circular in section whereas the lower booms consisted of wrought iron flat bars. The diagonal members were similarly of wrought iron in tension and cast iron in compression. The bearings which took the whole load of the bridge were at the top of the trusses and cast iron A-frames transferred the loads down to abutments.

The bridge was designed by Joseph Cubitt and erected

by Fox & Henderson in 1852, but it gained a reputation for lack of rigidity. In 1889 it was replaced by the present rivetted steel Whipple Murphy (double N) truss structure, one for each track as before, with a span of 262 ft and a depth of 26 ft at the centre of the curved upper booms, on the abutments of the previous structures. The Engineer was Richard Johnson, Chief Engineer of the GNR, and the contractor was Andrew Handyside of Derby.

The bridge was used in the 1920s for practical trials to determine, with the aid of mathematical analysis, the dynamic loads imposed on bridges by steam locomotives: the straight approaches on each side allowed high speeds to be attained, for test runs.

Further reading

CUBITT J. A. A description of the Newark Dyke bridge on the Great Northern Railway. *Proc. Instn Civ. Engrs*, 1853, **XII**, 601.

GRINLING C. H. *History of the Great Northern Railway*. Methuen & Co., London, 1898.

25. Fiddler's Elbow Bridge, Newark

HEW 53

SK 802 551

This is a striking, early reinforced concrete footbridge (originally a barge-horse bridge) across the Newark branch of the River Trent, ½mile upstream of the Newark Dyke railway bridge. Designed by L. G. Mouchel and

Below: Fiddler's Elbow Bridge, Newark, looking west

Partners and constructed in 1915, it is an arch of 90 ft span and 9 ft rise; the deck thickness is 10 in. at the springing and 6 in. at the crown. The deck rises steeply at a constant gradient from each end, with a sharp vertical curve at the apex — hence the name of the bridge.

The bridge is behind an industrial estate, away from the town streets, but it can be seen from the new A46 bypass on its south-east side, and also from the main line railway on its west side.

26. Flood Road, Newark (Smeaton's Arches)

HEW 1822

SK 794 546 to SK 788 561

The old Great North Road leaves Newark for the North by Newark Bridge (over the Newark branch of the River Trent) and runs straight for over 1 mile across the flood-plain as far as Muskham Bridge (over the Kelham branch of the river). John Smeaton designed this road on a low embankment penetrated by 74 arched passageways of 15 ft span in brickwork, which were built about 1770. The arches are in eight groups, the first group being immediately north of the A46 bypass, the last just before Muskham Bridge. The road over the arches is bounded by parapet walls with stone copings and there are cutwaters on the piers between them. The passageways were 33 ft long across the width of the road.

In 1929 they were extended in concrete by 13 ft on the west side to allow the road to be widened, but Smeaton's arches are all visible on the east side of the road. In 1990 Nottinghamshire County Council removed the road surface and filling from above some of the arches and replaced it with concrete, but the arches were left undisturbed.

27. Radcliffe-on-Trent Viaduct

HEW 741

SK 636 397

The viaduct carries the double-track Grantham–Nottingham railway line across the River Trent and is an interesting example of a cast-iron arch bridge that has been converted to a reinforced concrete bridge with little alteration to its general appearance.

It was built in 1848–50 by the Ambergate, Nottingham,

Boston and Eastern Junction Railway and the viaduct was its major engineering work.

The company had proposed four 50 ft spans across the river, but the Trent Navigation Company insisted on one span of at least 100 ft. In the event a 110 ft arch, cast by Clayton & Shuttleworth of Lincoln, was used across the north-western half of the river, the other half being spanned by three brick arches with stone voussoirs and abutments. The headroom above normal water level is about 24 ft; the arch rise is 11 ft 6 in.

The arch had six ribs, 3 ft deep in eight segments, the outer faces of the outside ribs having decorative features. Open-pattern spandrels supported the timber deck and the parapets are lattice pattern.

Timber approach viaducts were built across the flood plains on each side but that on the north-west side has been replaced by an earth bank and that on the other side by two runs of blue-brick arches, 28 in all.

In 1981 the iron ribs of the arch were encased in 28 in. wide concrete ribs with solid spandrels, but the outer iron ribs and spandrels have been left visible on the faces of the concrete. All the iron bracing and the timber deck have been replaced by concrete. This radical alteration, however, has been made inconspicuous by painting the exposed iron light grey and the concrete faces (where visible between the apertures of the spandrels) dark blue, so that the external appearance has been preserved, and the modification is noticeable only from close at hand.

28. Trent Bridge, Nottingham

HEW 724

SK 581 382

The name which many people associate with a cricket ground rather than with an historical engineering work is nevertheless an ancient one: a bridge across the Trent has been a feature of the south approach to Nottingham since the early 10th century.

The first bridge was built with stone piers and timber beams soon after the year 920. The second bridge, begun in 1156, had over 20 stone arches and a chapel at one end used by a religious fraternity who undertook to keep the bridge in repair. Nottingham Corporation took over the responsibility in 1551. Floods damaged the bridge from

Trent Bridge, Nottingham, west face

time to time and the northern half of the bridge was swept away by a flood in 1683.

The rebuilt or repaired bridge had fifteen pointed arches across the river and its washlands, giving openings totalling 347 ft in a total length of 538 ft, with a length of causeway at the south end, followed by two further flood arches.

By 1867 the foundations had become seriously scoured by the river and Nottingham Corporation instructed the Borough Engineer, Marriott Ogle Tarbotton, to design and build a replacement bridge. He chose a line just downstream of the old bridge and used coffer dams for the foundation work; the river bed was excavated about 5 ft to the underlying sandstone. Construction started in 1868 and was completed in 1871.

There are three main spans, each of 100 ft, comprising cast-iron arches of eight ribs braced by wrought iron girders. The face ribs are moulded and cast integrally with recessed spandrels, which have decorative gilded floral and foliage motifs. A moulded cornice runs the whole length. The parapets are of open ironwork and each archway has a central gaslight standard, now superseded by tall steel lighting columns. The width between papapets at the time of construction was 40 ft. The river

Shardlow Canal Port Clock Warehouse, Trent and Mersey Canal

piers are of ashlar, faced with clusters of polished Aberdeen granite columns and support seat recesses of decorative stonework. There are massive ashlar abutments, each surmounted by two great stone blocks on which are carved coats of arms. Outside these abutments are flood or footway arches, one on the north side, two on the south.

Benton & Woodiwiss of Derby was the general contractor and the iron work was made by Andrew Handyside & Co. of Derby. On completion of the new bridge, the old bridge was demolished, except for the two southernmost flood arches, which are preserved in the middle of a modern traffic island.

In 1924–26 the bridge was widened to 80 ft. The face ribs of the arches and all the face stonework on the downstream side were re-erected so as to maintain the original appearance. The original iron arch ribs were relieved by the insertion of reinforced concrete ribs in the spaces between them, and rivetted steel ribs were used for the widened part. The contractor was the Cleveland Bridge Engineering Company.

Upstream of the bridge for 1½ mile is a particularly pleasant stretch of the river, with stepped stone banks and a tree-lined drive backed by grassed areas.

Further reading

TARBOTTON M. D. *History of the old Trent Bridge with descriptive account of the new bridge.* Nottingham, 1871.

29. Trent and Mersey Canal: Shardlow Canal Port and River Dove Aqueduct

Trent and Mersey Canal: HEW 1135 SK 458 308 to SJ 567 810

Shardlow Canal Port: HEW 1067 SK 44 30

River Dove Aqueduct, Trent and Mersey Canal: HEW 1699 SK 269 269

The major part of this canal is included in the volume in this series on Wales and Western England (pp. 159 & 189). Work on the canal began in 1776 and the Derbyshire part was completed in 1770 . The canal runs from the River Trent at Wilden Ferry (SK 495 309 at the confluence of the River Derwent), to the Bridgewater Canal in Cheshire (which connected to the Mersey). It has broad locks as far as Burton-on-Trent and narrow locks thereafter. The Engineer was James Brindley, who was also working at the same time on the Chesterfield Canal.

Before the end of the 18th century, a canal port was established at Shardlow where the canal was crossed by the London–Manchester road (now the A6), 1 mile from the junction with the Trent. It had warehouses for goods being trans-shipped to or from narrow boats, or between canal boats and road transport, and buildings housing supporting trades like boat building, rope making, smithies and stables.

Shardlow is now a Conservation Area and still holds an impressive set of canal buildings, including several large warehouses with massive timber roof trusses, floor beams and columns. The best known is the four-storey Clock Warehouse built in 1780 which is 85 ft long and spans an arm of the canal with a 24 ft arch. It has recently been restored and converted into a restaurant; apart from plate glass windows the appearance has been preserved. A quarter-mile farther down the canal and on the same (north) side is another warehouse, newly restored for commercial purposes, which is probably the best preserved and the least altered of all of them.

Fourteen miles up from Shardlow is the canal's longest aqueduct carrying the canal across the River Dove near its confluence with the Trent, close to the Derbyshire–Staffordshire boundary. A row of twelve low, segmental brick arches, with wide piers between them, extends for 80 yd across the river and its flood plain. The aqueduct runs close to the old A38 road which crosses the Dove with an attractive bridge in classical style.

Further reading

SIVEWRIGHT W. J. *Civil engineering heritage: Wales and Western England.* Thomas Telford, London, 1986, 159–160, 189–192

30. Stapenhill Bridge, Staffordshire

HEW 1785

SK 253 219

In April 1889 an elegant suspension footbridge was opened across the River Trent, linking the village of Stapenhill on the south side of the river with the town of Burton-upon-Trent on the north bank. The bridge, which was a gift of Baron Burton, has a main suspended span of 120 ft with two side spans of 60 ft. The suspension cables are formed from flat wrought iron plates, three plates 8 in. wide being rivetted together at 4 in. centres to give a final thickness of 1½ in. Each individual plate is 18 ft long but the joints between adjacent plates are staggered so that one joint occurs every 6 ft along the cable.

The handrail of the bridge also acts as the deck truss, being of lattice construction with T-shaped and L-shaped members forming panels 6 ft. long and 5 ft 6 in. high. The trusses are attached to the suspension cables by wrought iron rods 1¼ ins in diameter at 6 ft centres. The width of the footway over the bridge is 11 ft 6 in. and the deck is

Stapenhill Bridge, Burton-on-Trent

formed of timber planks laid longitudinally over transverse lattice beams.

Below deck level the towers are formed from two 6 ft diameter cast-iron columns and above deck level they are square cast iron with decorative ribbing. An inscription on the tower records that the bridge was manufactured by Thornewill & Warham, Engineers, of Burton-on-Trent.

On the north bank, the footway is carried over the extensive flood plain of the River Trent by a 1673 ft long viaduct with 81 spans. Although the deck of the viaduct is modern the original 6 in. diameter cast-iron columns are still in use. Along the footway are two covered areas about 15 ft long with timber walls and a roof supported by iron roof trusses. These were evidently provided to give shelter to persons using the footway as it is very exposed to the weather.

EASTERN AND CENTRAL ENGLAND

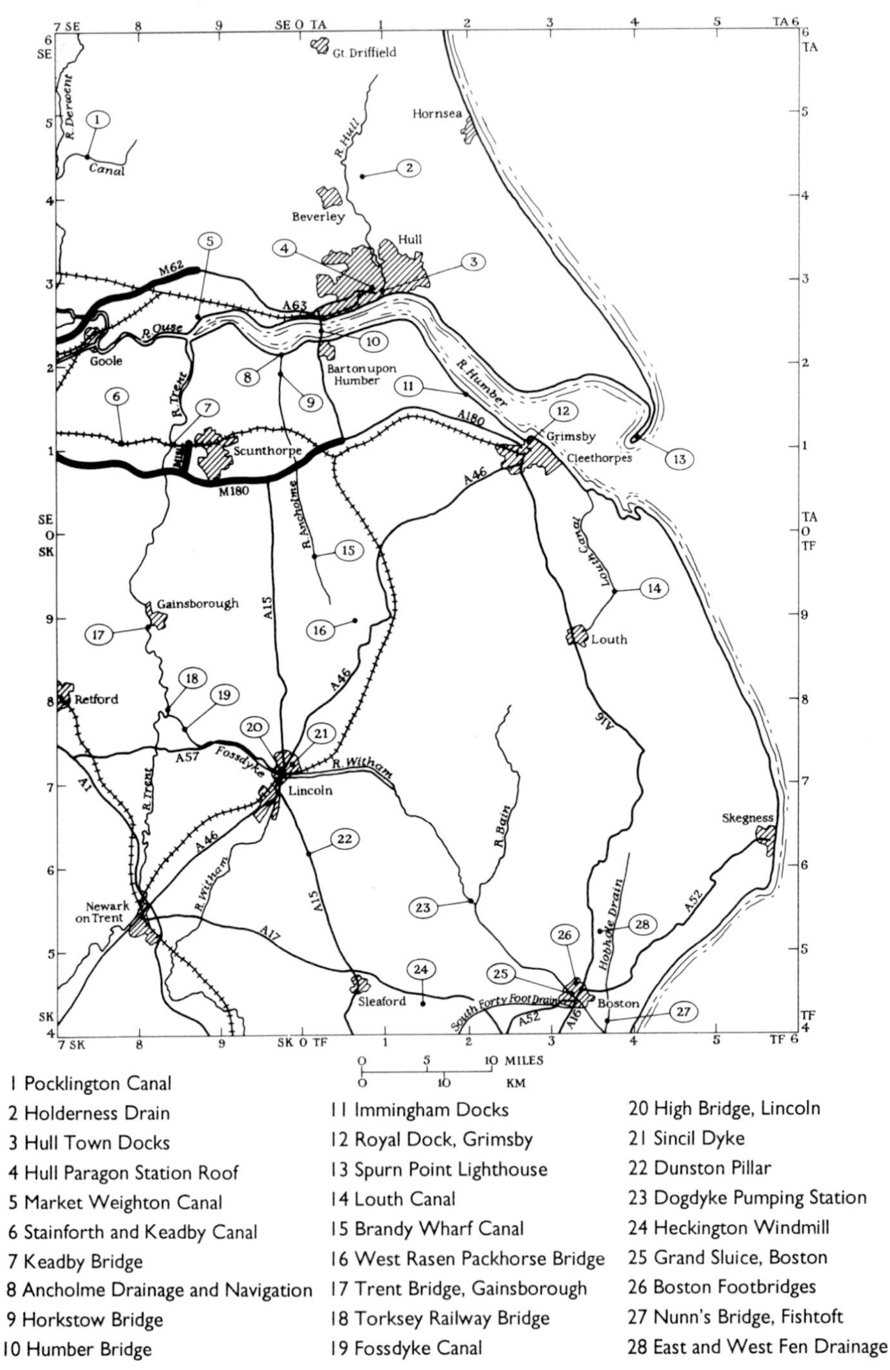

2. Humberside and North Lincolnshire

From the lower Trent and Ouse valleys, eastward to the North Sea, the topography of this region is dominated by the great ridges of chalk on either side of the Humber estuary, the Yorkshire and Lincolnshire Wolds. South of the Humber, the chalk scarp and dip landform is repeated to the west by the smaller oolitic limestone ridge, the two almost parallel ridges giving the North Lincolnshire landscape its characteristic appearance of ocean rollers set in stone.

Flanking these ridges are the lowlands of Holderness and the Lincolnshire Marsh, the carrs and fens of the Ancholme and Witham valleys, the Lincoln Clay Vale and the Isle of Axholme. Although land drainage and its associated engineering works are less of a major landscape feature of this region than they are further south in the Great Level of The Fens, drainage works are still a strong element of its civil engineering heritage.

From medieval times, large-scale intensive argiculture has been a notable activity in the region and, until the late 19th century, the region's commercial base was predominantly agriculture orientated, centered on Lincoln, Grantham and Gainsborough. Heavy industry, steelmaking in Scunthorpe and chemicals along the banks of the Humber, is relatively new to the area and has made a comparatively minor contribution to the engineering heritage of the region.

The primary lines of communications have always been on a north/south axis along the western edge of the region, the Ermine Street, the Great North Road and the Great Northern Railway. The eastern seaboard has had a less prominent, but economically vital, role as a maritime gateway to northern Europe. During the 13th century, Boston, with its wool trade, was the second largest port in the kingdom. Boston's trade declined but Hull, Grimsby, and latterly Immingham, still provide important outlets for the industrial North of England. The rich engineering heritage of the Humber area features in this section and the Humber suspension bridge takes pride of place as the most spectacular civil engineering structure in the region.

1. Pocklington Canal

HEW 827

SE 707 426 to SE 800 474

This canal links Pocklington, where the basin adjoins the York– Market Weighton turnpike road (now A1079), to the River Derwent and thereby to the Humber Estuary. In its length of 10 miles it also serviced the villages of Storthwaite and Melbourne. Its purpose was to provide a link for coal, lime, fertilisers and manufactured goods from the Humber with return loads of farming produce. It was plied by keels, which were 58 ft long, 14 ft 3 in. beam and had a draft of 6 ft 6 in.

The design was by George Leather and construction took place between 1814 and 1818. Use of the canal declined after the railway reached Pocklington in 1847 and it carried its last commercial traffic in 1932. It is being restored for pleasure craft used by the Pocklington Canal Amenity Society.

2. Holderness Drain

HEW 1947

TA 083 492 to TA 131 287

The Holderness Drainage District now covers 31 320 acres of low lying land bounded to the west by the banks of the River Hull and to the east by the higher Holderness Plain. This lowland area is at places 12 ft below high water levels in the Rivers Hull and Humber.

Up to the mid-18th century 11 200 acres of this low ground drained into the River Hull 4 miles north of its confluence with the Humber. The improvement of the drainage of this area was carried out under an Act of 1764 in the years 1765–67, to a design prepared by John Grundy. The work involved cutting a new main drain 1½ miles long from Foredyke and another from Ganstead, leading to Mother Drain, which was 1 mile long, discharging into the River Hull just above North Bridge. The works cost £24 000.

Under an Act in 1832 the drainage water was diverted to a new outfall to the Humber at Marflett 1¾ miles to the east of the confluence of the River Hull and Humber Estuary.

Further reading

The engineering works of John Grundy (1719–83). The Society for Lincolnshire History and Archaeology, 1985.

3. Hull Town Docks

HEW 81

TA 098 280 to TA 103 282

The natural sheltered harbour at Hull was initially on the River Hull extending about 600 yd upstream from its confluence with the Humber Estuary where the river was about 165 ft wide. Additional wharfage along the river was not possible, since this would have further restricted the flow in the river which was already subject to flooding.

Plans for a new dock were prepared by Henry Berry and modified by John Grundy. The dock was constructed in 1778 at a cost of £64 600. It enclosed a water area exceeding 10 acres and had an entrance lock 38 ft wide and 121 ft long with a 21 ft water depth. At the time of its construction it was the largest dock in England. Later, the dock was named Queen's Dock at the time of a visit by Queen Victoria.

The next stage was Humber Dock which was opened in 1809 and built at a cost of £233 000. The water area was 9½ acres with an entrance lock 42 ft wide, later called Prince's Dock, with a water area of 6 acres at a cost of £187 000.

A smaller dock, the Railway Dock, nearly 3 acres in area, was constructed in 1846 off Humber Dock. In the same year, a 50 ft long cast-iron swing bridge was built to carry a public road over the entrance lock to Humber Dock. The bridge now remains in a permanently fixed position.

Railway Dock and most of Humber Dock, which was shortened to allow the construction of the town bypass, are now busy marinas. Queen's Dock has been filled in and Princes Docks remains as a lake feature.

Further reading

1. BALDWIN M. W. The engineering of Hulls earliest docks. *Trans. Newcomen Soc.*, 1973–4, **46**, 1–12.

2. TIMPERLEY J. Harbour and docks at Kingston upon Hull. *Trans Instn. Civ. engrs*, 1836, 1–55.

4. Hull Paragon Station Roof

HEW 447

TA 092 288

The first station on the site was built in 1848 and was extended over the years as additional lines and railways entered Hull.

In 1904 the station comprised four platforms and three short bays; increased traffic necessitated remodelling to make nine full length platforms, a short bay and four excursion platforms. The new station was provided with a high roof forming a cover over all the lines and platforms in a manner common with major railway stations. The designer was William Bell.

The roof has five spans, varying from 62 to 72 ft, which are supported by four lines of thirteen cast-iron columns at 33 ft centres. The side spans are supported on station buildings and enclosing walls. There are roof trusses at each column and at three intermediate points on lattice girders which run between the column tops.

The trusses are arched with curved iron ribs tied at springing level. The lower part of the trusses is covered with timber boarding and the upper part glazed and raised above the truss members. The concourse across the station buildings end of the platforms is of a similar construction with two spans at right angles to the main station roof.

Further reading

BIDDLE G. and NOCK O. S. *The railway heritage of Britain*. Michael Joseph, London, 1983, 31, 34

5. Market Weighton Canal

HEW 213

SE 874 257

The Market Weighton Canal had a dual purpose as a navigable waterway for traffic between the Humber Estuary and Market Weighton and as an improvement to the River Foulness for better land drainage.

At the Humber end of the canal there is a tide sluice and a boat lock. The tide sluice consists of two masonry culverts through a masonry structure. The culverts are rectangular with a paved invert and have timber side hung gates on the Humber side which restrict the entry of water from the Humber into the canal. At low tide, drainage water in the canal spills over a weir and out through the gates.

The adjoining boat lock has four mitre gates; the outer two control the tide and canal water levels and the inner two gates form a lock for navigation for boats not exceeding 15 ft beam and 36 ft length.

The canal ceased to be used for commercial traffic in 1958 and was closed in 1971 but is still used by pleasure craft.

6. Stainforth and Keadby Canal

HEW 1758

SE 645 124 to SE 834 114

A 13-mile long canal was built between 1793 and 1802 to link the River Don Navigation at Stainforth to the Trent at Keadby. The route lies wholly through flat fen-like countryside and only one intermediate lock was needed between Stainforth Basin and the two-way tidal lock at Keadby. What distinguishes this canal from its contemporaries is its ambitious scale. Built to take 200 ton sea-going sailing vessels as far inland as Thorne, it was, in all but name, a ship canal.

The channel below Thorne is about 24 yd wide, except through bridges. All were built as swing bridges, though with the recent reconstruction of Medge Hall Bridge none are now of the original design. Perhaps the most intriguing is the horizontal rolling drawbridge near Keadby (SE 824 115) which carries the Doncaster–Scunthorpe railway across the canal at an acute angle, although this too is a relatively modern (1926) replacement. Still navig-

Opposite: Keadby Lock, Stainforth and Keadby Canal

able, the canal has no water supply of its own but relies on its end-on connection with the Don Navigation.

Further reading

HADFIELD C. *The Canals of Yorkshire and north east England.* David and Charles, Newton Abbot, 1972.

7. Keadby Bridge

HEW 847

SE 841 107

King George V Bridge, Keadby, opened in 1916, carries a double track railway and the A18, Doncaster to Scunthorpe road, over the River Trent 3 miles west of Scunthorpe. It was the Great Central Railway's biggest bridge undertaking and replaced their adjacent railway-only swing bridge of 1866, which was inadequate for the greatly increased traffic to their docks at Immingham.

This impressive steel bridge has three main spans: two fixed spans of 134 ft and 140 ft, and a lifting span giving a 150 ft clear waterway across the navigation channel. There are also two secondary spans on the eastern bank, a 70 ft approach span and a 40 ft span onto which the opening span rolls, giving a total distance between abutments of 548 ft. All five spans have three principal lattice

Keadby road and rail rolling lift bridge, looking north-east. Courtesy: *Grimsby Evening Telegrapoh*

girders, the central girder being located between the road and railway.

The structure was designed by James Ball and built by Sir William Arrol & Company. An American concept, the Scherzer rolling-lift bascule simultaneously rolls and rotates on its counterbalanced tail until the opening span is almost vertical. Keadby's 163 ft bascule was one of the first of its type in Britain and the heaviest in Europe. Until mains electricity became available the lifting mechanism was powered by storage batteries charged by petrol-driven generators. With the changing character of river traffic the bascule was fastened down permanently in 1960, when the roadway was widened and the headroom increased.

Further reading

BALL J. B. Keadby Bridge. *Min. Proc. Instn civ. engrs*, Part 1, 1916, 203, Nov., 33.

8. Ancholme Drainage and Navigation

HEW 844

TF 032 911 to SE 975 211

The River Ancholme flows northwards along a broad flat valley to enter the Humber estuary at South Ferriby. Below Bishopbridge the river was originally sinuous and tidal, the valley marshy and liable to flooding. Humber waters were first excluded by Sir John Monson's outfall sluice near Ferriby in 1640, no doubt influenced by Vermuyden's drainage at Hatfield Chase across the Trent, but this structure failed. Material for the sluice had been taken from the ruins of Thornton Abbey and its collapse ascribed to 'a just judgement of God'. Although it was replaced circa 1679 the drainage of the valley remained unsatisfactory, probably because the sill of the new sluice had been set too high. In these schemes, attempts had also been made to improve the river itself.

In his 1801 report, John Rennie recommended catchwater drains down each side of the valley, discharging to the Humber through separate sluices. Some work was carried out but the scheme was not completed until Rennie's son, Sir John, implemented a modified version of his father's proposals between 1825 and 1844. The river was comprehensively straightened, widened and bridged for the 19 miles to Bishopbridge, Brigg was bypassed and a single intermediate lock erected at Harlam Hill. Sir John Rennie also built the present Ferriby Sluice, a substantial masonry structure with three 18 ft waterways and roadway above, incorporating both a navigation lock and the western catchwater outfall.

Further reading

RENNIE Sir John. An account of the drainage of the Level of Ancholme, Lincolnshire. *Min. Proc. Instn Civ. Engrs*, 1845, 4, 7.

PAGE C. J. *History of the Ancholme Navigation*. Lincolnshire Local History Soc., 1969.

9. Horkstow Bridge

HEW 598

SE 973 190

A fine example of an early suspension bridge, Horkstow Bridge is one of only a handfull in Britain to have remained as originally designed. Horkstow Bridge, by virtue of its secluded location at the end of a little used country lane, has survived almost intact from the time it was completed in 1836.

Situated 1 mile upstream of Ferriby Sluice, it was built as part of Sir John Rennie's River Ancholme Drainage Scheme of 1825–1844 (HEW 844). The imposing and well proportioned rusticated masonry towers, which rise to a height of 36 ft above river level, have semi-elliptical arches over the 14 ft wide roadway. The single 133 ft 9 in. span has double wrought iron suspension chains, one immediately above the other, on either side of the bridge. Each chain consists of two 3/4 in. by 1 in. links, 7 ft 3 in. long. There are 33 pairs of 7/8 in. square iron hangers from which are slung the timber cross members supporting the simple plank deck. The deck has a pronounced upward convexity and its lightweight superstructure results in a significant motion under live loads, even with a 2 ton weight limit.

In September 1979 an articulated lorry weighing 23 tons attempted to cross the bridge. The tractor unit fell through the timber deck into the river and became detached from its trailer. The driver was lucky to have escaped unhurt as his cab was completely submerged. One quarter length of the bridge deck was destroyed and the suspenders in this section were bent outwards at their lower ends, one being broken. The masonry arches, chains and the remaining three quarters of the deck survived without damage. The bridge was subsequently repaired.

Further reading

LEWIS M. J. T. Horkstow Bridge. Lincs. *Ind. Archaeol.* **8**, 1, 1973.

10. Humber Bridge

HEW 204

TA 024 246

Opened to traffic in June 1981 after eight years under construction, this superb modern suspension bridge was, at the time, the longest span of any bridge in the world. The main span of 1410 m and the asymmetric north and south side spans of 280 m and 530 m respectively are suspended from hollow reinforced concrete towers that reach 155.5 m above the tops of their piers, the first major suspension bridge to use concrete in this way. Each 700 mm diameter, 5500 tonne suspension cables consists of 37 strands, each strand having 404 individual 5 mm high tensile steel wires. The welded steel deck structure is 28.5 m wide overall and 4.5 m deep and is suspended from the main cables by inclined hangers.

The deck carries a dual carriageway road with a combined footpath/cycle track on each side. The bridge was designed by Freeman Fox & Partners. John Howard & Co. built the towers and substructure and a consortium, British Bridge Builders Ltd., erected the steel substructure.

Further reading

WILKINSON G. *Bridging the Humber.* Cerialis Press, York, 1981.

11. Immingham Dock

HEW 1949

TA 195 162

Under the chairmanship of Sir Edward Watkin, the Great Central Railway's period of expansion did not end with its new main line to London. Physically and commercially constrained at Grimsby, the railway company needed its own deep water port on the east coast. Immingham was the obvious choice, within Great Central territory and where the deep water channel came closest to the Humber bank.

The Humber Commercial Railway & Dock Company,

a Great Central creation, obtained an Act of Parliament in 1901 for a dock at Grimsby but this was amended in 1904 for the greenfield site at Immingham. Designed by Sir John Wolfe Barry and constructed between 1906 and 1912 by Price, Wills & Reeves at a cost of £2.6 million, it represented a major example of a comprehensive Edwardian engineering project.

The single 45 acre basin with 1¾ miles of wharfage, designed to handle both bulk and general cargoes as well as passenger traffic, is set within a 1000 acre estate served by 170 miles of dock railway and sidings. Entrance to the dock is through an 840 ft x 90 ft lock flanked by river walls and approach jetties and equipped with three pairs of gates. The development included seven hydraulic coal hoists, transit sheds, granary, offices and a 740 ft x 56 ft graving dock. Construction of the dock was heavily mechanised, using massive steam excavators; 320 000 yd^3 concrete and 30 000 yd^3 of brickwork were used, with Swedish granite for the coping and corner stones. The whole complex was opened by King George V and Queen Mary on 22 July 1912.

Further reading

DOW G. *Great Central*. **3**, Ian Allan, London, 1963 229–236, 242–249, 259–262.

12. Royal Docks, Grimsby

HEW 776

TA 277 110

Grimsby Docks date from 1800, although the tidal creek where the River Freshney entered the Humber estuary had for centuries been a small harbour for coastal shipping. The first dock was created by dredging the creek and constructing an entrance lock across its mouth. The Grimsby Haven Company entrusted the work to a local builder who was unable to cope with the severe foundation problems that he encounted. John Rennie was then brought in as consulting engineer and under his guidance the 145 ft x 37 ft lock was successfully completed in 1800. It was closed in 1917 and the chamber subsequently infilled, although the inner end and wing walls are still visible.

The coming of the railways to Grimsby initiated the next and most important phase in the development of

'The construction of the Royal Docks, Grimsby', J. W. Carmichael, circa 1850, Institution of Civil Engineers

Grimsby's docks. The Royal Dock was located north-east of the Old Dock but constructed wholly outside the Humber Bank, projecting out into the estuary so that the entrance locks were in deep water and capable of taking the largest ships of their day. The Royal Dock was built inside a massive cofferdam extending ½ mile into the estuary and enclosing 138 acres. It was erected in 1846/8 and consisted of a 1 mile long earth embankment, incorporating wharfage for offloading materials and, in the lock area, 500 cu. yd of clay-fill between timber piling, parts of which can still be seen at low tide.

Because of poor ground conditions the Engineer, James Rendel, adopted the same unusual design for his quays which Rennie had employed on a much smaller scale for his lock 50 years previously. Behind the 8 ft thick masonry quay walls lie a series of semicircular brick arches spanning 33 ft between brick piers on piled foundations, like a concealed viaduct, extending back up to 72 ft from the quayside. The 32 ft high quay wall is pierced by smaller round-headed openings below water level to equalise pressure between the dock and the vaults behind.

The Royal Dock, which is 2200 ft long and 500 ft wide, was opened by Prince Albert in May 1852. The twin parallel entrance locks were 300 x 70 ft and 200 x 45 ft respectively, with 27 ft depth of water on the cill at spring tides.

Between the entrance locks rises the elegant and distinctive Dock Tower by J. W. Wild, inspired by Siena's medieval town hall. The prominent 309 ft brick tower built in 1852 contains an elevated water tank providing sufficient pressure to operate the lock gates and dockside cranes, one of William Armstrong's earliest applications of hydraulic power. Long disused, it is believed to be the only hydraulic systems of its type to be built. Development of Grimsby's Docks did not end with the Royal Dock and continued until 1934 when Fish Dock No. 3 was completed, bringing the total enclosed water area to 137 acres.

Further reading

NEATE C. Description of the cofferdam at Great Grimsby, *Min. Proc. Instn Civ. Engrs*, London, 1849–50, 91, No 813 1849/50.

CLARKE E. H. Description of the Great Grimsby (Royal) Dock. *Min. Procs Instn Civ. Engrs*, London, 1864/5, 24,38.

13. Spurn Point Lighthouse

HEW 296

TA 403 112

The Spurn is a spit of sand and gravel about 3½ miles long, the result of littoral drift from the coast of Holderness, which protects the Humber Estuary and is subject to periodic erosion and regrowth.

The first lighthouse on Spurn Point was built in 1427 and employed a coal fire as its light source. This was followed by a number of structures with very limited life until Smeaton designed a tower, built in 1771–76, which was 90 ft high. The lighthouse was necessarily founded on sand and eventually had to be shored up by timber frames.

Smeaton's lighthouse was demolished in 1895 and was replaced by the present lighthouse designed by Thomas Matthews of Trinty House. The lighthouse is 120 ft high and is constructed of Staffordshire blue brick, the walls being 4 ft 10 in. thick at the base and 2 ft 9 in. at the top.

Further reading

1. HARWOOD B. Spurn Point and its lighthouses. *The Engineer*, **80**, 1895, Oct. 4, 344.

14. Louth Canal

HEW 1948

TF 338 879 to TF 354 032

A geographically isolated canal this was built to give the town of Louth a navigable communication with the sea for Yorkshire keels. Taking its water supply from the chalk fed River Lud at Louth, it follows an 11½ mile north easterly route, serving also as a main drain, through the flat coastal plain. From the sea lock at Tetney to the Riverhead basin at Louth all but one of its eight locks come in the last 4 miles. Surveyed by John Grundy in 1756 and opened in 1770, it was one of the first canals of the canal age.

Its principal point of interest is the unusual design of the upper six locks. The brick walls of the timber floored lock chambers are scalloped in plan with four segmental arches each side, concave to the chamber. Vertical timber posts at the intersections of the arches are connected by tie bars to ground anchors to resist lateral earth pressures. Only one other lock of this design is known to have been built elsewhere and that is also in Lincolnshire. The navigation was abandoned in 1924 and is now derelict, though some progress has been made to preserve Grundy's unique locks. The lower end of the canal has, since 1968, been used to convey water to Covenham Reservoir.

Further reading

SKEMPTON A. W. *The engineering works of John Grundy (1719–1783)* Soc. for Lincs. History and Archaeology, 1985.

15. Brandy Wharf Bridge

HEW 1712

TF 015 970

This neat single-span, cast-iron road bridge over the new channel of the River Ancholme was built by Sir John Rennie in 1831, in connection with his Ancholme drainage and navigation scheme (HEW 844). The five cast-iron ribs, which form the 54 ft 9 in. span, each consist of two half-span castings springing from masonry abutments and bolted together at the crown. The ribs, cast by the Butterley Company, are interconnected by spacer tubes, tie-rods and diagonal cross-bracing.

At some time (when, so the story goes, a large experimental trenching machine was taken across the bridge during World War I) the arch was distorted, re-

Brandy Wharf Bridge, following strengthening

sulting in a small but permanent deformation. In 1988 the bridge was sensitively strengthened and refurbished by Lincolnshire County Council, who inserted four slightly arched steel beams in the gaps between the cast- iron ribs and replaced the original cast-iron plate decking with a composite reinforced concrete deck slab. The original cast-iron arch, still perceptibly deformed, now carries no live load. The adjoining wharf and warehouse are of the same period.

Further reading

BOYES J. and RUSSELL R. *The canals of eastern England*. David and Charles, Newton Abbott, 1977, 295–96.

16. West Rasen Packhorse Bridge

HEW 914

TF 063 494

In the centre of this tiny village 2½ miles west of Market Rasen, in a county somewhat lacking in ancient bridges, a medieval bridge in miniature spans the River Rase. It is reputed to have been built by Bishop Dalderby in the early 14th century though it is probably of later date, post Black Death, when the wool trade rose to prominence.

The three segmental ribbed sandstone arches are of roughly equal span (11 ft to 12 ft), giving a total length between abutments of 39 ft. The cobbled trackway over the bridge is fairly constant 4 ft 6 in. wide between 2 ft high parapet walls which, with the very modest rise from bank to crown, would have been ideally suited to laden packhorses. The sandstone has weathered considerably over the centuries but this delightful little bridge is still in reasonable condition. It is a scheduled Ancient Monument.

Further reading

JERVOISE E. *The ancient bridges of mid and eastern England.* Architectural Press, London, 1932, 56.

West Rasen Packhorse Bridge, looking north-east

17. Trent Bridge, Gainsborough

HEW 1774

SK 815 891

Gainsborough has a handsome and generously proportioned ashlar masonry bridge across the River Trent. When completed in 1791 it was the only bridge across the river below Newark, although there appears to have been a short lived structure at Knaith, 2 miles upstream, in the 12th century. Its three 26 ft 3 in. wide semi-ellipitical arches have 62 ft, 70 ft and 62 ft spans respectively. The engineer was William Weston, little known in Britain, as shortly afterwards he emigrated to America, where he had a distinguished engineering career. Trent Bridge was originally promoted as a toll-bridge and remained so until 1932; the toll houses still flank the carriageway at the eastern end of the bridge. It was a commercial success from the start. The bridge carries the heavily trafficked A631 road and was widened in 1964. This was achieved, somewhat to the detriment of its appearance, by the addition of cantilevered concrete footways in place of the stonework balustrades. Despite its age and exposure to modern traffic, the structure shows no apparent signs of deterioration or distress.

Further reading

PEVSNER N. and HARRIS J. *The buildings of England: Lincolnshire.* (2nd ed) Penguin Books, London, 1989.

18. Torksey Railway Bridge

HEW 198

SK 836 792

Torksey railway bridge, built in 1849, carried the Manchester, Sheffield & Lincolnshire Railway's double track Retford to Lincoln line across the River Trent. The bridge has twin 130 ft river spans and a 570 ft long eastern approach viaduct. As originally built, the two main spans comprised a pair of tubular wrought iron girders, rectangular in sections, 10 ft high and 2 ft 3 in. wide. Wrought iron cross members, level with the base of the main girders, supported the bridge decking and rail tracks at 2 ft intervals.

Right: Torksey Lock, looking north-east, Fossdyke Canal

John Fowler's design was bold and innovative, similar to Robert Stephenson's contemporaneous Britannia Bridge over the Menai Strait (HEW 110). At first, the Commissioners of Railways refused permission for the bridge to be opened, criticising both design and workmanship, but in April 1850, after four months of technical argument, the Commissioners finally authorised the opening.

Increased traffic loading eventually necessitated the strengthening of the bridge and in 1897 a conventional steel lattice girder was inserted centrally into each main span, the original southern girders being moved outwards to maintain track clearances. The line across the Trent was closed in 1959. Further downstream, Gainsborough Railway Bridge (HEW 846, SK 810 882) of 1848/9 was also by Fowler and very similar in design to Torksey. It has recently been rebuilt.

Further reading

FAIRBURN W. On tubular girder bridges. *Proc. Instn, Civ. Engrs.* 1849–50, 9. 223–287.

19. Fossdyke Canal

HEW 513

SK 834 781 to SK 970 713

The evidence for the Fossdyke's Roman origins, though circumstantial, is compelling and, since the original purpose of the Car Dyke (HEW 822) is uncertain, the Fossdyke is probably the only Roman canal still extant in

Britain. The western portion of the Fossdyke appears to be largely artificial whereas the eastern end is possibly a canalisation of the River Till. It was reopened to navigation during the reign of Henry I and remained in intermittent use for the next six centuries.

The present eleven mile long waterway, linking the Trent at Torksey with the River Witham Navigation at the Brayford Pool in Lincoln, is approximately 80 ft wide and 5 ft draught. It dates from the 1740/45 reconstruction by Richard Ellison. There is only one lock, at Torksey, which has four pairs of gates as, at times of flood, the level of water in the Trent can exceed that of the canal.

The Fossdyke relies for its water supply on inflows from the River Witham and Till and its level is controlled by weirs and sluices on the Witham at Stamp End Lock, Lincoln. Barge traffic between the Fossdyke and lower Witham has always been severely hampered by the limited draft and headroom under Lincoln's High Bridge (HEW 845).

Further reading

WHITWELL B. *Roman Lincolnshire.* The Society for Lincolnshire History and Archaeology. 1970

BOYES J. and RUSSELL R. *The canals of eastern England.* David and Charles, Newton Abbot, 1977, 254–268.

20. High Bridge, Lincoln

HEW 845

SK 975 711

The High Bridge has carried Lincoln's High Street over the River Witham since about 1160 and incorporates the ribs of what is believed to be the second oldest masonry arch bridge in Britain, the oldest being at Fountains Abbey in Yorkshire. It is certainly the only bridge in the country with a medieval secular building still standing on it.

The oldest section of High Bridge is about 33 ft long with a single 22 ft span and consisted originally of semicircular stone barrel vaulting with five ribs, the outer pair 4 ft wide and the three inner ribs 3 ft wide. Around 1235, a 28 ft long extension, with a 22 ft span, was built to accommodate a chapel. This extension had quadripartite groined stone vaulting with diagonal ribs. At some later date two of the intermediate ribs of the original arch were

Lincoln High Bridge, East End, looking upstream

removed, as was much of its stone vaulting, together with the vaulting between the ribs of the 13th century extension. This was replaced by brickwork, presumably to lower the roadway, although the 1902 restoration partially reinstated the original layout. High Bridge was extended upstream by 20 ft in 1540–50 when a twin-centred flat pointed arch without vaulting was constructed to support a range of half-timbered buildings which, now restored, are still in commercial use. A minor downstream extension was added in 1762/3, bringing the total length of the bridge to 87 ft.

The medieval chapel, disused since 1549, was demolished by 1762 and a stone conduit house in the form of an obelisk was erected in its place, but in 1939 this too was demolished, being an obstruction to motor vehicles. Wooden flooring along the river bed beneath the bridge was removed between 1792 and 1795, the bed lowered and the abutments underpinned, to facilitate the passage of barges, although the bridge remained a serious obstacle to flow and navigation. High Bridge continued to carry the main road until 1971 when a relief road was opened and High Street became a pedestrian precinct. The bridge is a scheduled Ancient Monument and Grade I Listed Building.

21. Sincil Dyke

HEW 1814

SK 969 696 to SK 983 708

Sincil Dyke, a 1½ mile long artificial channel through Lincoln's industrial heartland, is the least known of Lincoln's waterways. Its origins are obscure but it is almost certainly pre-medieval and probably Roman. It serves as a flood flow relief channel from Bargate Weir on the River Witham, discharging into John Rennie's South Delph on the eastern edge of the city thus bypassing the constricted river channel through the city centre. Prior to Rennie's early 19th century Witham Drainage scheme it rejoined the river at Stamp End.

The Sincil Dyke's original purpose is unkown. It is possible that it formed part of a defensive work running parallel to the Ermine Street causeway, now Lincoln's High Street, but it has certainly had a primary drainage function since the mid 13th century. Hydraulically, the Witham, the Fossdyke Canal navigation (HEW 513) and Sincil Dyke have always been intimately linked, with the low-lying area west of Lincoln, draining to the Sincil Dyke by means of a culvert under the river at Great Gowt since circa 1792.

The smaller Great and Little Gowt Drains, which intersect the High Street to link the Witham and Sincil Dyke are themselves of considerable antiquity. In 1847, part of the channel was culverted in connection with the development of the Great Northern Railway's Central Station and with progressive urbanisation much of the channel is now confined within retaining walls.

Further reading

WHEELER W. H. *History of the fens of South Lincolnshire.* (2nd ed), Boston and London, 1896.

22. Dunston Pillar

HEW 1024

TF 009 620

Dunston Pillar, almost certainly Britain's only land lighthouse, was erected in 1751 by a local landowner, Sir Francis Dashwood, to guide travellers across the desolate and uninhabited expanse of heathland south of Lincoln. As originally built, the 92 ft high tapering square stone tower was surmounted by a lantern, and at its base Dashwood provided refreshment rooms and pleasure gardens. The lantern was lit regularly until 1788 when enclosure of the heath began to make the lighthouse redundant. It was used for the last time in 1808 and, in 1810, the Earl of Buckingham subsitituted a statue of George III for the lantern. During World War II the statue was removed and the tower truncated to minimise the risk to aircraft. The stump of the tower stands just off the A15 Lincoln to Sleaford road, 6 miles south of Lincoln, and the bust of the statue is now in the grounds of Lincoln Castle.

Further reading

PEVSNER N. et al. *The buildings of England — Lincolnshire.* (2nd ed), Penguin Books, Harmondsworth, 1964, 520.

23. Dogdyke Pumping Station

HEW 774

TF 206 558

This is a typical small land drainage pumping station of its time or, to be more precise, a pair of adjoining pumping stations, each characteristic of its own era. Pumped fen drainage on this site began in 1796 with the erection of a wind pump, a 36 ft sail driving a 16 ft water-wheel, to lift water from the Mill Drain into the River Witham. This was replaced in 1855 by a Bradley & Craven low pressure, double acting 16 hp A-frame rotative beam engine with 12 ft 3 in. beam, 14 ft flywheel and separate condenser. It drove a 24 ft diameter scoop wheel at about 7 rpm, draining an area of around 1500 acres.

In 1940, the steam pump was superseded by a 40 hp Ruston & Hornsby diesel engine driving a 22 in. Gwynnes centrifugal pump. The old pumping station was later abandoned and its chimney, a landmark for enemy aircraft, demolished. The diesel pump was subsequently replaced by twin automatic electrically driven pumps, although the diesel installation is still retained by the Internal Drainage Board for emergency use.

Unfortunately for continuity, the steam and diesel pumps lie directly under the flight path of a nearby military airfield and the new station is sited a quarter of a mile downstream, outside the aircraft noise hazard area. A preservation trust was formed in 1973 which has restored the steam engine to working order albeit with a modern boiler. Both the steam and diesel engined stations are now part of a common Scheduled Ancient Monument.

Further reading

Dogdke pumping stations Preservation Trust. *Lincolnshire Life*, June 1978, 43.

WHEELER W. H. *History of the fens of South Lincolnshire*. 2nd ed., Boston and London, 1896, 196.

24. Heckington Windmill

HEW 691

TF 145 435

Heckington Mill, five miles east of Sleaford, was built in 1830 as a five-sailed tower mill, but in 1890 the original sails and cap were wrecked by a sudden tailwind and the mill was abandoned. Fortuitously, a Mr Pocklington, who had bought at auction the mechanism, cap and sails of Boston's eight-sailed Skirbeck Mill of 1813, was looking for a suitable location to install his purchase. By 1892 Heckington Mill was back in operation with eight sails.

John Pocklington died in 1941. By 1945 his fine mill had again fallen into disuse but in 1953 it was acquired by the then Kesteven County Council, in order to preserve it. The mill is now owned by Lincolnshire County Council who carried out considerable restoration work. It is the last remaining eight sailed windmill in Britain; only seven or eight of which were ever built. It has a five storey brick tower with ogee cap, fantail and eight single-sided patent sails. The tips of the sails are, possibly uniquely, tied together by iron rods. The eight sails generated considerable power, even in light winds, and from 1892 the mill operated five pairs of millstones and an adjoining sawmill.

Heckington is one of a number of fine tower mills to be seen in east Lincolnshire; also notable are Boston's recently restored Maud Foster Mill (TF 332 447), Dobson's Mill at Burgh le Marsh (TF 503 649) the five-sailed Alford Mill (TF 457 766) and the six-sailed Sibsey Trader Mill (TF 344 510), a late example of its kind (1877, by Saunderson of Louth) but as superb a display of the millwright's art as can be found anywhere in Europe.

Further reading

DOLMAN P. *Lincolnshire windwills — a contemporary survey*. Lincs. County Council, 1986.

Heckington Windmill, looking west

25. Grand Sluice, Boston

HEW 1872

TF 324 445

The origins of the Grand Sluice go back as far as 1142 when a 'great sluice' was erected across the River Witham at Boston, though by 1316 this wooden structure was 'ruinous and in great decay'. The river became tidal again until, in 1500 a Dutchman, May Hake, built a masonry and timber structure across the narrowest part of the river, just upstream of the present Town Bridge. Although the sluice itself functioned barely a hundred years, the bridge above the sluice lasted until 1807, spectacularly decrepit in its later years. Hake's sluice consisted of an asymmetrical pair of waterways, 44 ft and 21 ft wide respectively, separated by an off-centre pier. The two waterways were closed against the tide by horizontally hinged flap doors suspended from iron hooks, a dubious design for doors of this size as experience proved.

The present Grand Sluice, by Langley Edwards of Kings Lynn, was completed in 1766 and is possibly the earliest tidal outfall sluice still extant in a substantially original state. A plain and rather squat masonry structure carrying a four brick arch roadway, it has three 17 ft wide waterways and an integral navigation lock, enlarged in 1881, on the east side, which can be used as a relief waterway in times of flood. As originally built, it had timber pointing doors on both the seaward and landward sides but the latter were replaced by vertical steel lifting gates in 1979/82.

Further reading

WHEELER W. H. *A history of the fens of south Lincolnshire*. 2nd ed. Boston and London, 1896.

26. Boston Footbridges: Cowbridge Footbridge and Hospital Lane Footbridge

Cowbridge footbridge: HEW 643 TF 328 471

Hospital Lane Footbridge: HEW 644

John Rennie's earliest cast-iron bridge, Town Bridge in Boston, was built in 1803 to carry the main street across the River Witham. This bridge was replaced in 1913, but Rennie also erected three other cast-iron bridges in and

Boston Footbridge

around the town. A trio of almost identical footbridges spanned Rennie's newly enlarged Maud Foster Drain, the high level outfall to his East and West Fen drainage system. Two of them, at Cowbridge and Hospital Lane, are still extant. Pevsner calls them elegant, as indeed they are.

The single arch of each bridge is formed by a pair of slender cast-iron ribs giving a 61 ft clear span with a 3 ft 9 in. rise. The ribs, which bear the inscription 'Cast at Butterley 1811', are also curved in plan. This gives the two bridges a markedly waisted appearance and illustrates the skill of the early ironfounders. The webs of the ribs are pierced in the Vierendeel style and their upper chords are bowed to give a pronounced camber to the footways. Simple wrought iron balustrades terminate in plain square gritstone pillars above brick abutments.

The third of the trio, Vauxhall Bridge, ½ mile south of Hospital Lane, was demolished and replaced by a road bridge in 1924.

Further reading

WRIGHT N. R. The iron bridges of Boston, *Linc. Ind. Archaeology*, 1972, 7, 1.

27. Nunn's Bridge, Fishtoft

HEW 1743

TF 367 415

To the casual observer this is just an ordinary modern concrete bridge carrying a minor road across the Hobhole Drain, three miles south east of Boston. The structure has a clear span of 72 ft between abutments and an overall deck width of 20 ft. The five main prestressed concrete beams, interlinked by five equally spaced reinforced concrete stiffeners, support an integral deck slab. Built in 1947/8 to replace an earlier brick structure, Nunn's Bridge was constructed by the Witham Fourth District Internal Drainage Board, using direct labour under their Engineer G. E. Buchner.

What makes Nunn's Bridge of such significance is that it was the first prestressed concrete bridge to be built *in situ* in Britain. The five 3 ft 7 in. deep by 1 ft 3 in. wide I-section main beams are each reinforced with twelve prestressing cables arranged in a complex pattern of catenaries. Each 1¼in. diameter post-tensioned cable comprises twelve ⅕ in. diameter high tensile wires spaced round the circumference of a wire helix, the whole being spirally wrapped and sheathed with a 5 in. wide steel strip. The design was by L. G. Mouchel & Partners, although the Drainage Board and its Engineer deserve credit for their boldness in adopting such a pioneering design. The structure is still in excellent condition.

Further reading

STANLEY C. C. *Highlights in the history of concrete*. Cement and Concrete Association, 1979.

28. East and West Fen Drainage

HEW 1950

TF 210 554 and TF 471 604 to TF 336 431

The East and West Fens, a 92 sq. mile tract lying between Boston and the Wolds to the north, were the last major area of fen to be drained in England. Successive piecemeal attempts had been made to drain the area, notably by Sir Anthony Thomas in the 1630s, but by the start of the 19th century large parts of these fens, particularly the East Fen, were still what would today be regarded as an extensive wetland supporting a subsistence economy of summer grazing, wildfowling and fishing.

In 1801 Sir Joseph Banks of Revesby promoted an Act

of Parliament to drain the whole area, including Wildmore Fen and engaged John Rennie as Engineer. Hitherto the area had outfalls to the River Witham at Antons Gowt above Boston, through the Maud Foster Sluice in Boston and to the Steeping River above Wainfleet. Rennie's typically thorough solution was to construct catchwaters round the wold margins of each fen, bringing the upland waters together into an enlarged Maud Foster Drain to a new Maud Foster Sluice. Provision was made for the lowland waters of the West Fen to be conveyed beneath the catchwater at Cowbridge and into the East Fen. Here they were discharged into a completeley new channel, the 13 mile long Hobhole Drain which, with various artificial tributaries, now drained the East Fen, taking the lowland waters to a completely new outfall sluice on the Witham estuary four miles below Boston. The whole project was complete by 1807. Its layout remains substantially unchanged to the present time, although the inevitable shrinkage of the peat in the East Fen necessitated the construction of an intermediate pumping station on the Hobhole Drain at Lade Bank in 1867.

Further reading

WHEELER W. H. *A history of the fens of south Lincolnshire.* 2nd ed, Boston and London, 1896.

Key to map for section 3

1 Car Dyke
2 Grimsthorpe Park Dam
3 Lolham Bridges, Maxey
4 Stamford Canal
5 Wansford Bridge
6 Ferry Bridge, Milton Park
7 Nene Railway Bridge, Peterborough
8 Triangular Bridge, Crowland
9 Deeping Gate Bridge
10 Horseshoe Bridge, Spalding Common
11 Fenland Drainage
12 Pinchbeck Marsh Pumping Station
13 Sea Bank, The Wash
14 River Nene Outfall
15 Cross Keys Bridge, Sutton Bridge
16 North Level Outfall Sluice, Foul Anchor
17 Morton's Leam
18 Denver Sluice
19 Bedford Rivers, Old and New
20 Ely Cathedral, Octagon Lantern
21 Stretham Pumping Station
22 St Ives Bridge
23 Houghton Mill
24 Huntingdon Bridge
25 Chinese Bridge, Godmanchester
26 Bourn Windmill
27 Mathematical Bridge, Cambridge
28 Clare Bridge, Cambridge
29 Garret Hostel Bridge, Cambridge
30 Trinity College Bridge, Cambridge
31 St John's College Bridges, Cambridge
32 Magdalene Bridge, Cambridge
33 Wicken Fen Windpump
34 Hobson's Conduit, Cambridge
35 London to Cambridge Railway

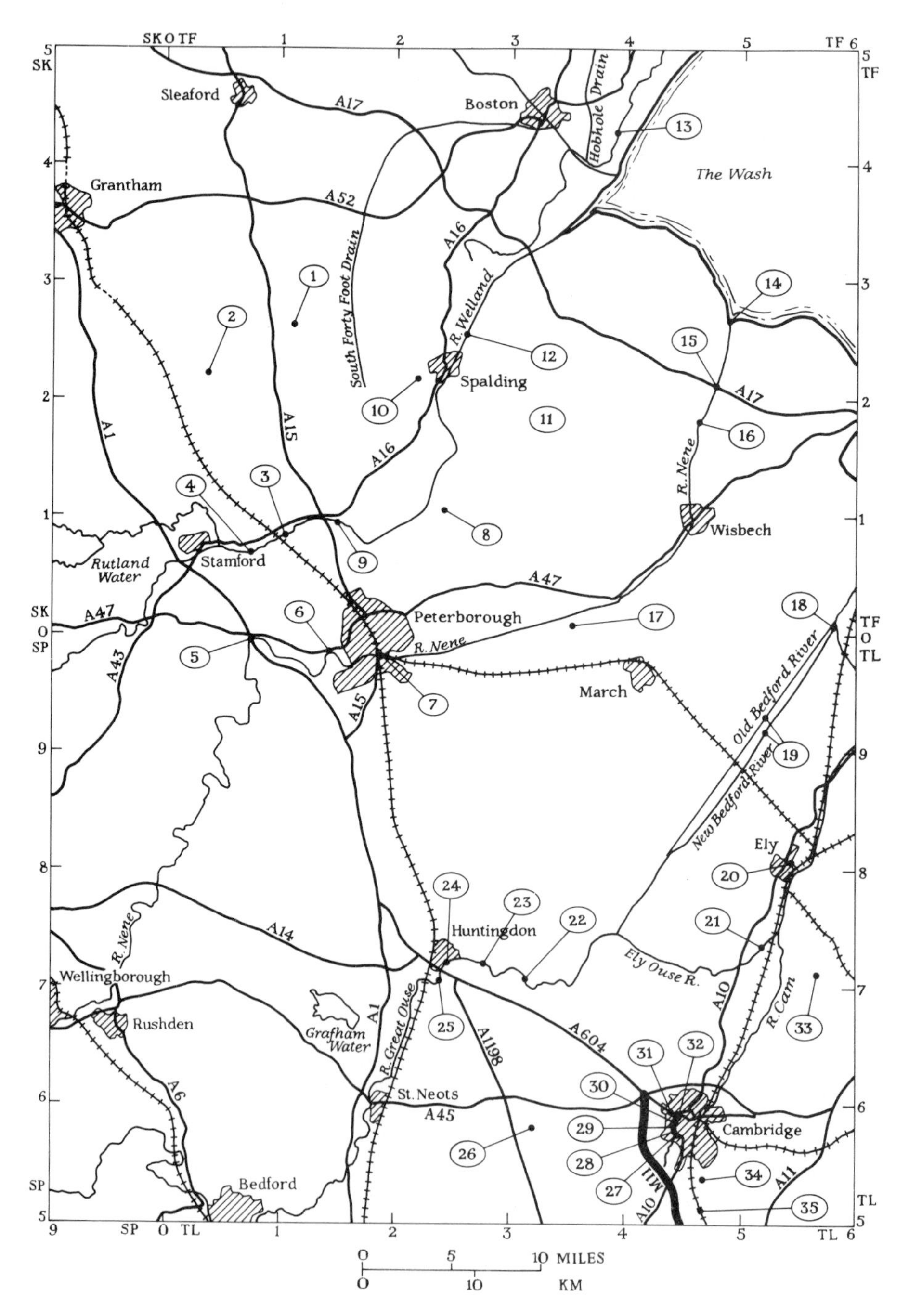
Sleaford
Boston
Grantham
The Wash
Hobhole Drain
South Forty Foot Drain
R. Welland
Spalding
Stamford
Rutland Water
Peterborough
R. Nene
Wisbech
March
Old Bedford River
New Bedford River
Ely
Huntingdon
Ely Ouse R.
R. Cam
Wellingborough
Rushden
Grafham Water
R. Great Ouse
St. Neots
Cambridge
Bedford
A17
A52
A16
A15
A1
A47
A43
A14
A604
A1198
A45
A6
A10
A11
M11
0 5 10 MILES
0 10 KM

3. South Lincolnshire and Cambridgeshire

The sweeping landscape of the Fens today owes much to the work of civil engineers over the centuries. Formerly, large areas of South Lincolnshire and north Cambridgeshire lay seasonally, or even permanently, under water. The land was known to be valuable and worth reclaiming but the task was immense. The problem was to carry the runoff from the adjacent upland country across the low-lying land to reach the sea and, at the same time, to prevent the incursion of tidal waters. The medieval period saw some realignment of the rivers to deal with flooding caused by reduced out-flow and, in the 15th century, there were further attempts to treat the effects of siltation, by straightening the river courses.

The first co-ordinated drainage schemes came in the 17th century with the formation of a group, called the Adventurers, who began to reclaim the Great Levels. At the forefront of this enterprise was the Dutch engineer, Vermuyden, and the inheritance of his great drainage system is still evident across Fenland. At the same time, the potential of windmills to raise water from low-lying lands into the embanked rivers was recognised. Unfortunately, the peat lands shrank as they dried out and drainage became less effective. This problem became acute as more efficient steam engines began to replace the windmills from the beginning of the 19th century and it was necessary to deepen the outfalls of the main rivers at The Wash. In turn, the internal drains were lowered and the cycle of land shrinkage continued. The drainage system still requires constant attention and the National Rivers Authority is engaged in a continuing programme of improvement.

As railways penetrated the region, local industries began to prosper and in Peterborough extensive engineering works developed. The town also became a main railway centre, but remains a market location for a wide surrounding area. The early impact of railways on the other main centres, Cambridge and Ely, was less pronounced, each preserving its individual character. Cambridge has a rich heritage of noble buildings and attractive bridges and Ely has its cathedral with the great lantern of the Fens.

1. Car Dyke

HEW 822

TF 05 69 to TL 20 98

The origin of this waterway has long been the subject of considerable research and debate. It was originally thought to have been a navigable waterway, but it now appears that would not have been possible and it is suggested that it was constructed by the Romans about the year AD 130 as a catchwater drain. It was carefully laid out to run along the foot of the Lincolnshire Wolds from the River Witham at Lincoln to the River Nene at Peterborough, crossing the Rivers Slea, Glen and Welland in its course. In an area where changes in level are quite small, some skill was needed to arrange a continual fall from each watershed to take the upland waters away and Car Dyke may be seen as the first large scale effort to improve the Fenland.

The dyke has silted up over the years, and in places been recut, so its original dimensions are not certain. Excavation suggests a width of 45 ft and depth of 13 ft for the Lincolnshire section, which compares favourably with many of the canals of the 18th century. Despite the centuries of neglect, its course is visible over long stretches.

Further reading

TEW D. H. Roman waterways in the Fens. *Trans. of the Newcomen Soc.*, **52**, 1980–81, 139–149.

Top: Car Dyke at Potterhanworth Booths

2. Grimsthorpe Park Dam

HEW 542

TF 039 219

Grimsthorpe Castle, the home of the Earls of Ancaster, lies in attractive rolling landscape 3 miles north west of Bourne. In the extensive parkland behind the castle there are two ornamental lakes, the lower of which, Great Water, has an earth dam which was designed by John Grundy of Spalding and completed in 1748. The embankment is 420 ft long and has a maximum height of 18 ft.

The outstanding feature of this dam is the pioneering use by Grundy of a rammed clay corewall. Probably about 6 ft wide, it extends the full length and height of the embankment and down into a cut-off trench beneath the embankment in order to prevent leakage.

Although Grundy returned to Grimsthorpe Park in 1758 to deal with leakage through swallow holes in the bed of Great Water, the dam itself proved to be watertight and survives intact.

Further reading

SKEMPTON A. W. *The engineering works of John Grundy (1719–1783).* **19**, The Society for Lincolnshire History and Archaeology, 1984.

Grimsthorpe Park dam of the Great Water

3. Lolham Bridges, Maxey

HEW 1762

TF 111 069 to TF 111 074

The Roman road from London to Lincoln crossed the River Nene at Durobrivae and continued north-west on the higher ground past Stamford. From the Nene another Roman road, King Street, struck due north, crossing the Welland valley where it is lower and wider and going on to join the Ermine Street at Ancaster. The Welland crossing probably remained in use through Saxon and later times so that new bridges were required to cater for the increase of wheeled traffic in the 17th century.

The present Lolham Bridges consist of fourteen arches of moderate span, up to 16 ft, grouped in five structures over the several channels of the river. The arches are of 17th and 18th century origins and were built by the County of Northampton. The County authorities were generally reluctant to spend money in this way, preferring to lay the responsibility on the landowners or parishes, but in this case they left records of their munificence in the form of carved stone tablets on the cutwaters of the bridges.

Further reading

The Victoria county history of the counties of England — Northamptonshire. **2**, Univ. of London, 1906, reprinted 1970, 502.

Lolham Bridges

4. Stamford Canal

HEW 1945

TF 031 069 to TF 164 090

Despite being one of Britain's oldest post-Roman canals, surprisingly little is known about the Stamford Canal. An Act of Parliament of 1571 authorised the restoration of an earlier river navigation up the Welland, utilising the old river channel or contiguous mill-streams. Nothing appears to have happened until 1620, when a Commission of Sewers confirmed Stamford Corporation's right to make a 9½ mile cut, with twelve locks and associated sluices, from Market Deeping to Stamford. Even then it seems that the bulk of the work was actually carried out much later, between 1664 and 1673, by which date the canal was in use.

Although the lowest two miles from Deeping Lower Lock to Market Deeping Mill and the final ¾ mile from Hudd's Mill to Stamford Wharf are river navigations, the intermediate 6½ mile cut running parallel to the river is a proper canal. The 1¾ mile section to West Deeping utilised an existing mill-stream but the remainder of the cut was undoubtably purpose built.

Proposals in the early years of the 19th century to link the Stamford to the Oakham and Grand Union Canals, creating a through route from the Wash to the Midlands, came to nothing. The coming of the railway to Stamford in 1846 dealt the canal a fatal blow and it fell into disuse in 1863. Much of the line of the canal can still be followed and the masonry remains of the two river locks on the Welland at Deeping Gate and Deeping St Nicholas, the former a 60 ft by 12 ft pound lock and the latter, possibly a turf-walled lock, are easily accessible.

Further reading

BOYES J. and RUSSELL R. *The canals of eastern England.* David & Charles, Newton Abbot, 1977, 234–241.

RUSSELL R. *Lost canals and waterways of Britain.* Sphere books, 1983.

5. Wansford Bridge

HEW 51

TL 074 992

This bridge, where even today two cars cannot pass, carried the Great North Road over the River Nene, until the road was diverted in 1930.

The earliest reference comes from 1221 when the

bridge was already old enough to need repair and the Bishop of Lincoln, in whose diocese it lay, granted an indulgence to those who would contribute. In 1234 it was replaced with oak donated by the King. In the next century it became the responsibility of the people of the village of Wansford, who were frequently granted the right to levy pontage (bridge tolls) to pay for its upkeep. By 1586 its importance to the wider travelling public was recognised when it was ordered to be repaired by the County.

The twelve-arch structure is of three dates. The seven northern arches date from the late 16th century rebuilding. Originally they all had double arch rings in two orders but the fourth to the seventh arches have been strengthened with a brick vault and now have triple arch rings. A stone inscribed 'P M 1577' is placed in a refuge south-east of the seventh arch, opposite a marker of the old county boundary. The next three arches were rebuilt after floods in 1672–74 and are considerably higher than their earlier neighbours. They have single arch rings only and spans up to 25 ft 6 in.

After another flood had swept away two arches over the main river the opportunity was taken to improve the navigation of the river and one arch of 50 ft span was built in their place. Long-and-short voissoirs and a datestone of 1795 mark this work. Lastly, a small arch on the south bank which may be 17th century, was widened on the west side when the main arch was rebuilt. The bridge is slightly over 600 ft long and the 16th and 17th century structure is 14 ft wide between parapets.

It was bypassed in 1930 by another noteworthy structure, a span of 109 ft, designed by Sir Owen Williams, which is claimed to be the largest mass concrete arch on the British road system. It has cast into it the names of two local authorities, County of Huntingdon and Soke of Peterborough, and now carries northbound traffic on the rerouted A1 road.

6. Ferry Bridge, Milton Park

HEW 1763

TL 143 984

A tablet on the face of this bridge records that it was 'built at the sole cost and charge of Rt. Hon. William, Earl

Fitzwilliam. 1716'. He was the third baron and had been raised to the rank of earl that year. The Fitzwilliams lived at Milton Hall, immediately to the north of the Gunwade Ferry which this bridge replaced; the bridge provided access over the Rver Nene to the south without the need to go to Peterborough or Wansford. It has never been a public road and when Daniel Defoe went to cross it in a carriage about the year 1724, he was charged 2/6 of which he would

> 'only say this, that I think 'tis the only half crown toll in Britain'. (Daniel Defoe. *A tour through the whole island of Great Britain*. PRC 1989, 151.)

This very pleasing structure has three semicircular arches which spring well above normal water level. Triangular cutwaters and fluted keystones provide relief to the fine ashlar limestone. The north abutment contains two small rooms, with doors on the east side and small bull's-eye windows on the west.

Reconstruction of the trunk road A47 in 1991 has meant the removal and rebuilding of the Ferry Lodge further to the north-west, and Ferry Bridge now gives access to the recreational areas of Peterborough's Nene Park.

Ferry Bridge, Milton Park

7. Nene Railway Bridge, Peterborough

HEW 93

TL 191 981

The Great Northern Railway (HEW 1921, section 6) crossed the Rivers Great Ouse and Nene by cast-iron bridges. The former, near Huntingdon, was rebuilt in steel in the 1930s but the latter survives, with only a little strengthening, to carry the two up tracks of the East Coast Main Line.

The use of cast iron for medium-span railway underbridges was fairly commonplace by the late 1840s and Brassey, the contractor for this stretch of the line, was building several with spans of up to 120 ft. At Peterborough, the River Nene is crossed by three arches, each of 66 ft span and 8 ft 6 in. rise. There are six ribs, set in pairs at 3 ft centres, each pair being at 7 ft 6 in. centres. The ribs themselves are in two segments bolted together at mid-span, so that they could be erected with a minimum of disruption to the navigation below. The river piers consist of twelve cast-iron fluted piles, two below each rib. The spandrels have diagonal lattices supporting the bridge deck.

The piers and the ribs were strengthened early in the 20th century by steel bracing bolted or clamped to the cast

Nene Railway Bridge, Peterborough

iron, and the spandrels have been reinforced. This bridge is a scarce survivor of its kind — others are at Dewsbury (HEW 347 & 348) and Shrewsbury (HEW 903).

When this section of the route was widened to four tracks, a Whipple Murphy truss bridge was built, parallel to the cast-iron bridge, to carry the two down lines.

Further reading

COSSEY F. Cast iron railway bridge at Peterborough. *Industrial Archaeology*, 1967, 4, May, 138–147.

8. Triangular Bridge, Crowland

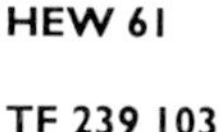

HEW 61

TF 239 103

This unique bridge was built in the centre of Crowland to span the confluence of the River Welland and its tributary, the Cattewater or Cat's Water. Built of timber in the 10th century, the bridge was rebuilt in the late 14th century using limestone brought from Ancaster, 25 miles away. The river system of the Fens has been altered out of all recognition since that time and now the Welland is channelled 1 mile to the west of the town and Cat's Water is culverted locally.

The structure consists of three half spans, set at 120° to each other in plan, meeting at the apex of the bridge. Each

Triangular Bridge, Crowland, looking north

half span is 10 ft long and is supported by three ribs. The external ribs are intricately moulded, showing the influence of the masons of Croyland Abbey who constructed the bridge. It is doubtful whether it was ever of much practical use, being too steep for any except foot traffic, and it has been suggested quite plausibly that it was intended to be a parable in stone on the nature of the Trinity.

Further reading

JERVOISE E. *Ancient bridges of mid and eastern England*. The Architectural Press, 1932, 67–68.

9. Deeping Gate Bridge

HEW 63

TF 150 095

This fine three-arch bridge over the River Welland is dated on its south-west cutwater, 1651, a time when the English Civil War had just ended but political uncertainty still prevailed.

In engineering terms, the spans are conservative and the arch shape is nearly semicircular. An ashlar band at road level, double arch rings in two orders, each with a small chamfer, and a deep label over each arch give a degree of architectural distinction.

Deeping Gate Bridge, upstream elevation

The bridge has spans of 18 ft, 19 ft, and 18 ft 6 in. Triangular cutwaters are carried up to provide refuges at road level, a very necessary feature as the roadway is only 13 ft wide between parapets and there is no footpath. The structure is generally of limestone although the central arch vault has been rebuilt with engineering brick.

Further reading

JERVOISE E. *Ancient bridges of mid and eastern England.* The Architectural Press, 1932, 67.

10. Horseshoe Bridge, Spalding Common

HEW 1832

TF 219 211

The Horseshoe Bridge is a single 45 ft span reinforced concrete road bridge across the South Drove Drain 2 miles south-west of Spalding; its superficially modern appearance belies its age and historical significance. It is one of a pair of such bridges built in 1910 for the Deeping Fen Trustees, for which the Liverpool Ferro-Concrete Company won the contract in direct competition with conventional steel beam designs. Like most early reinforced concrete bridges they were designed by L. G. Mouchel & Partners using the French 'Hennebique' system. The entire structure, possibly even the piled foundation, was built in concrete.

The working drawings, which still exist, are of considerable interest; the reinforcement detailing is, in many respects, archaic but the basic design concepts are clearly established. The structural members were all cast *in situ* and the steel reinforcement is continuous throughout, with no apparent construction joints.

The structural arrangement is somewhat unusual. The parapet walls form the two principal longitudinal beams and are linked below deck level by nine transverse beams spanning between them. This layout, although giving the 22 ft wide deck a lightweight appearance, means that the bridge is vulnerable to impact damage, especially from modern lorries, and it now carries a 7.5 tonne weight limit. Horseshoe Bridge's twin, Money Bridge on the River Glen near Pinchbeck, was demolished in the mid-1980s.

11. Fenland Drainage

HEW 83

TF 16 to TL 67

The Fenland covers a significant portion of three counties: Cambridgeshire, Lincolnshire and the western side of Norfolk. It is generally grade 1 agricultural land, much of it the most fertile in the country. Large parts of the area lie below sea- level and the wetlands have been the scene of an increasing contest, stretching over many centuries, between drainage engineers and the forces of nature.

The great marshes fringing the Wash were originally flooded for most of the time. Draining the area presented a two-fold task, to control the flow of water coming from higher ground inland, across the fens, and at the same time, to prevent the sea from surging across the marshes at high tide.

Early attempts to control and improve drainage were made in the 15th century, notably the making of Morton's Leam (HEW 823), a direct channel to relieve the flow in the River Nene, east of Peterborough. At that time, the church took the lead in maintaining land drainage but, with the dissolution of the monasteries in the 16th century, this initiative halted.

It was not until the 17th century that a full-scale effort was launched to control the river courses across the marshes. The fourth Earl of Bedford, with the support of a group of Adventurers, cut a straight channel from Earith to Salters Lode, now known as the Old Bedford River, to divert water from the River Great Ouse.

The master stroke came in 1651 with the construction of the parallel New Bedford River by the 5th Earl, creating a large area between the two cuts which could be used to impound flood water. Both river projects, and many other associated schemes, were planned and directed by the Dutch drainage engineer, Vermuyden, whose influence has extended to the present day. An essential feature of his main river scheme was the construction of a sluice at Denver (HEW 821), to hold back tidal waters and to allow the rivers to discharge on ebb tides. More detailed descriptions of certain drainage works, including the Bedford Rivers and Denver Sluice, are to be found in this section.

As the Fenland drainage continued during the 18th and 19th centuries the reduction of water content in the soil led to land shrinkage and erosion. The land level fell below that of the main rivers, and wind pumps (as many as 750) were used to lift water from the dykes into the rivers. In the 20th century, diesel and electric pumps have replaced the wind pumps. Land shrinkage continues and the Black Fens have become an upside-down world with the embanked rivers above the surrounding land. Parts of the fen country are now totally dependent on the complex drainage and pumping systems which have been evolved over the centuries, as well as the tidal defences.

Further reading

DARBY H. C. *The draining of the Fens*. CUP, 1940.

12. Pinchbeck Marsh Pumping Station

HEW 1711

TF 262 261

Pinchbeck Marsh has the distinction of being the last beam engine and scoop wheel to work in the fens. The station drained an area of 4000 acres, between Vernatt's Drain and the River Glen, from 1833 to 1952, when its boiler became unserviceable. A 20 hp low pressure A-frame rotative beam engine with a 14 ft beam and 18 ft 6 in. flywheel, believed to have been built by the Butterley Company, drove a 22 ft diameter scoop wheel at 6¾ rpm. It is still housed in its original brick pumphouse, although the chimney has gone, and, apart from the substitution of a piston valve for its slide valve mechanism in 1919, the engine is also in its original state.

The steam engine was replaced by an adjacent electrically driven pumping station in 1952 missing out the intermediate diesel engine era completely. The old pumping station was, however, retained intact by the Welland and Deepings Internal Drainage Board until 1988 when, in partnership with the South Holland District Council, they restored the beam engine and scoop wheel to running order (not in steam) as the centrepiece of a small but fascinating museum of land drainage.

Further reading

CROWLEY T. E. *The beam engine*. Senecio Press, 1982.

WHEELER, W. H. *A history of the fens of south Lincolnshire*. (2nd edn.) Boston & London, 1896.

13. Sea Bank, The Wash

HEW 1946

TF 53 59 to TF 60 21

The Sea Bank was an earthen embankment which created the shoreline of The Wash and kept the sea from the low-lying land behind. It stretched from Wainfleet near Skegness to north of King's Lynn, making large loops inland towards Spalding and Wisbech, and covered a distance of some 45 miles. There is no record of its construction, but a map of the fenland settlements mentioned in the Domesday Book suggests very strongly that it existed more or less in its entirety by then. The 17th century historian Dugdale called it 'Roman Bank' but there is no evidence that it was constructed by them. Quite possibly it was a series of local efforts though the continuity of line suggests that there was a good degree of co-operation.

Substantial lengths of the Sea Bank still exist today, some as much as 10 miles from the present shore line. In places it is 10 ft high, though interestingly the surface of land on the seaward side is significantly above that inland. This is not due to shrinkage, as the soil here is not peat, but to the deposition of silt by the tides through the medieval period. Embankments constructed in the last two centuries have consolidated the intake of these new lands, leaving the Sea Bank as a reminder of the early struggle to guard against the ravages of the sea.

Further reading

DARBY H. C. *The changing fenland*. CUP, 1983, 7–10.

14. River Nene Outfall

HEW 1274

TF 468 182 to TF 495 278

The importance of a clear and adequate outfall where the River Nene flows into The Wash had long been known to play a vital part in the drainage of the northern Fens. In 1773, Kinderley's Cut was made from Wisbech to Gunthorpe Sluice to improve the outfall, but the value of this new channel was progressively negated by silting of the river at Cross Keys Wash.

At the beginning of the 19th century the consulting engineer John Rennie drew up plans to cut a new channel 200 ft wide and 5 miles long, close to the former bank of Cross Keys Wash. The work was carried out between

1827 and 1830 under the direction of Thomas Telford, who had been appointed as Rennie's successor. The contractors, Jolliffe & Banks, employed up to 1500 men to make the cut. On completion, the old river was closed off and the full discharge from the Nene was then sufficient to scour out the new channel to the required depth for

River Nene Outfall

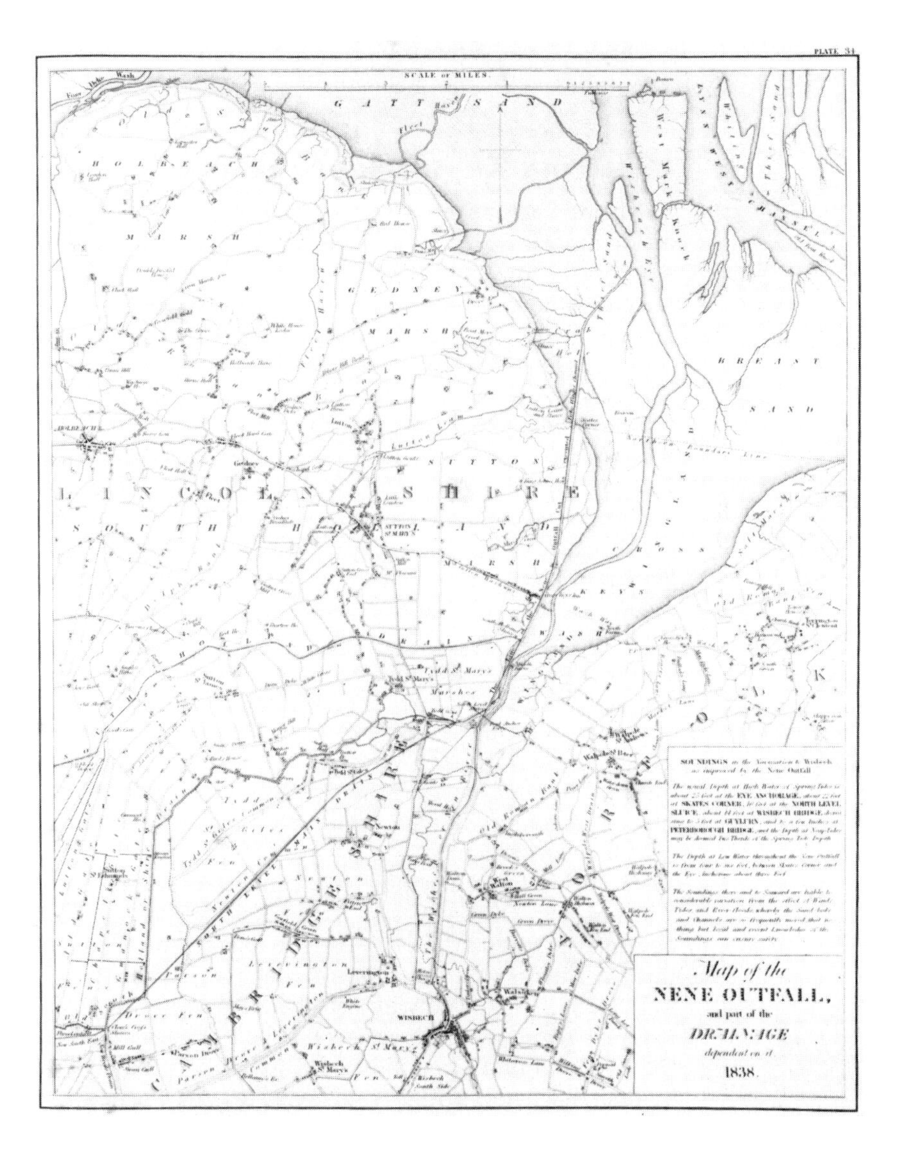

navigation within a few months. The banks of the new cut were protected with 10 000 tons of stone pitching.

The Nene outfall Amendment Act of 1829 authorised the River Commissioners to erect two lighthouses at the mouth of the river, without the sanction of Trinity House. The brick towers are 40 ft high with small cottages attached and are now privately owned. The towers have never displayed lights but served as landmarks for shipping entering the outfall channel.

15. Cross Keys Bridge, Sutton Bridge

HEW 1275

TF 482 210

The bridge across the River Nene at Sutton Bridge, just inside Lincolnshire, has a strange history. Beginning as a road bridge, it was converted to the dual use of road and rail and later reverted to road-only use. Because of the flat terrain, the limited height which can be obtained for the river crossing has always required an opening span to allow a river passage for shipping.

The initial road bridge, built by Sir John Rennie in 1825, was 650 ft long, consisting of 14 timber spans and a cast-iron opening span of 52 ft. The bascule mechanism was operated by hand, which must have been an arduous task because of the frequent passage of ships to and from Wisbech. Twenty-five years later, it became necessary to reconstruct the bridge because of changes in the river channel.

The replacement bridge was built by Robert Stephenson in 1850, to a more elaborate design than the earlier structure. A swing span of 193 ft pivotted at the centre on a pier in the river. The wrought iron plate girders, 8 ft deep at the pivot tapering to 4 ft deep at the nose ends, turned on an 8 ft diameter roller path. There were three side spans of 52 to 60 ft.

The bridge was destined to be converted to dual road and rail use. The Lynn and Sutton Bridge Railway Co. obtained powers in 1864 to use the bridge and a single line track across the south side of the bridge was opened in 1866. By 1890 the railway stretched from Peterborough to Great Yarmouth and was soon to be incorporated into the Midland and Great Northern Joint Railway. It was evident that a new bridge was needed to carry the heavier locomotives then in service and powers were obtained to construct it about 100 ft south of the old bridge and to move the road and railway on to the new alignment. Joint road/rail use was still required as it was unacceptable to have two bridges close together across a shipping channel.

The third bridge, designed by J. Allen McDonald of the Midland Railway, was completed in 1897. It has a swing span of 166 ft with an off-centre pivot and two approach spans of 50 ft and 77 ft. The main span comprises three steel trussed girders, 16 ft deep at the pivot, with the centre girder between the road and railway. The bridge swung by hydraulic power and, for the first time at this site, the control cabin was built on top of the swing span. The steelwork was fabricated by the Staffordshire Steel Co. of Bilston and erected by the Derby firm of Andrew Handyside. The total cost was £80 000.

The railway closed in 1959 and four years later the bridge was converted back to sole road use. It is a Listed Grade 2* structure.

Further reading

Clarke R. H. *History of the Midland and Great Northern Joint Railway.* Goose & Son, Norwich, 1967.

The Engineer, 23 July 1897, 79-80.

16. North Level Outfall Sluice, Foul Anchor

HEW 1764

TF 467 181

The present North Level outfall sluice is the fourth to be built, each on a different site, since the first was constructed as part of Vermuyden's works under an Act of Parliament of 1649. That first sluice did not last long and was soon replaced by a structure nearer the river.

The diversion of the River Nene (HEW 1274) to a new channel and the cutting of the new straight North Level Main Drain to replace the old winding Shire Drain, necessitated the building of a sluice at the new junction. The old sluice was left isolated on the east bank of the river, where a farm is still called Gunthorpe Sluice Farm. The new sluice which was designed by Thomas Telford, had a cill 8 ft lower than the old one and the works led to a quick and dramatic improvement in the drainage of the level.

After a period of time, improved runoff brought about by steam pumping, together with the scouring of the River Nene's bed, led to the North Level Act of 1857 which authorised the widening and deepening of the North Level Main Drain and the construction of a sluice with a cill 5 ft 4 in. lower still, a little to the north. On this occasion the design was provided by George Robert Stephenson and it was built by Smith and Knight. A plaque records its construction in 1859. Three openings of 10 ft, 20 ft, and 10 ft were provided, with rising oak gates upstream and hinged iron gates downstream to keep out the tides. The site of the 1830 sluice is still clearly visible, at a chicane in the road to Foul Anchor.

Almost immediately there were problems with seepage. When the wing walls bulged in 1866, they were taken down, rebuilt and a cast-iron strut placed between to hold them in position; this doubles as a footbridge. Later repairs are also recorded on the downstream face.

A couple of disastrous sluice failures in the Middle Level in 1862 had left everyone very nervous about safety. The North Level Commissioners decided to build another sluice a little upstream as a belt and braces measure. The work was carried out in 1871 without interruption to the flow in the drain, by an ingenious use of caissons. B. P. Stockman was in charge of the construction.

Further reading

MILLER S. H. and SKERTCHLY S. B. J. *The Fenland past and present.* Longmans Green & co, 1878, 165.

17. Morton's Leam

HEW 823

TL 208 974 to TF 397 030

In the medieval period, the Fens were marshy for much of the year and subject to widespread and destructive flooding in winter from the drainage of the uplands along the courses of the main river outflows to The Wash. Only the islands of higher ground such as Peterborough and Ely were relatively immune from the risks of being engulfed in periodic floods. The perceived solution was to improve the flow in the main water courses, both in capacity and gradient, but only the great abbeys had the resources and incentive at that time to tackle the problem.

An early effort to improve land drainage was the cutting of Morton's Leam, so called after Bishop John Morton of Ely, who devised a scheme to bypass the winding course of the River Nene by the construction of a direct channel between Stanground, near Peterborough and Guyhirne. The 12 mile cut, as originally dug in the period from 1478 to 1490, was 40 ft wide and 4 ft deep. It was the first channel in the Fens to use, on a large scale, the practice of making artificial rivers in straight cuts as a means of improving hydraulic gradients.

After the dissolution of the monasteries the Leam became neglected and partially blocked, a condition which was corrected in the 17th century by the work of Cornelius Vermuyden and the development of Fen drainage schemes on a major scale. Morton's Leam still serves as a useful element in the Fen drainage network.

Further reading

DARBY H. C. *The Medieval Fenland*. CUP, 1940.

DARBY H. C. *The draining of the Fens*. CUP, 1940.

18. Denver Sluice, Norfolk

HEW 821

TF 587 010

The success of Cornelius Vermuyden's grand design for the drainage of the fenlands depended on a means of preventing the ingress of tidal water into the River Great Ouse system and the retention of sufficient river water for navigation. This was achieved in 1652 by the construction of a sluice at Denver where the waters of his Old and New Bedford Rivers (HEW 820) rejoined the Ouse. The great sluice, coupled with the sluice up the river at Earith, now named Hermitage Lock, effectively separated the flows originating in Bedfordshire from those of Cambridgeshire.

Denver Sluice was wrecked in 1713 by a combination of a river flood and a surge tide and, in the years which followed, the South Level of the Fens was again open to inundation. This sorry condition was remedied when the sluice was rebuilt in 1746–50 by a Swiss engineer named Labelye. The sluice that he constructed lasted until the present sluice, the third on this site, was designed and built by Sir John Rennie in 1834. It had three sluiceways, each 17 ft 6 in. wide with cills 6 ft lower than the 1750 Labelye design. The Rennie structure also incorporated a lock. This structure still survives but the upstream timber sluice gates have been replaced by steel lifting gates powered by electric motors. A new 'eye', or sluiceway, was added by the Ouse Drainage Board in 1923 to improve the discharge capacity of the sluice under flood conditions. It is 35 ft wide with a cill at 10 ft below Ordnance Datum.

A further improvement in flood control at Denver was provided by the Great Ouse Flood Protection Scheme which was commissioned in 1965. A relief channel was cut from Denver to King's Lynn with sluice gates at both ends which enabled flood waters from the South Level rivers to bypass Denver and to be discharged instead at King's Lynn. Here, the low water level is some 11 ft below that at Denver. To ease the flood level in the Ely-Ouse a cut off channel, 30 miles long, was made from the River Lark near Mildenhall, crossing the River Little Ouse and the River Wissey, to take flood waters from all three rivers to Denver for discharge into the Relief Channel. The concept of the cut-off channel had been foreseen by Vermuyden in his scheme of 1638 and, with the completion of the Protection Scheme over 300 years later, his dream of flood control in the entire South Level system was realised.

Further reading

DARBY H. C. *The draining of the Fens.* CUP, 1940.

WELLS S. *The history of the draining of the Great Level of the Fens called Bedford Level,* Parts 1 & 2. 1830.

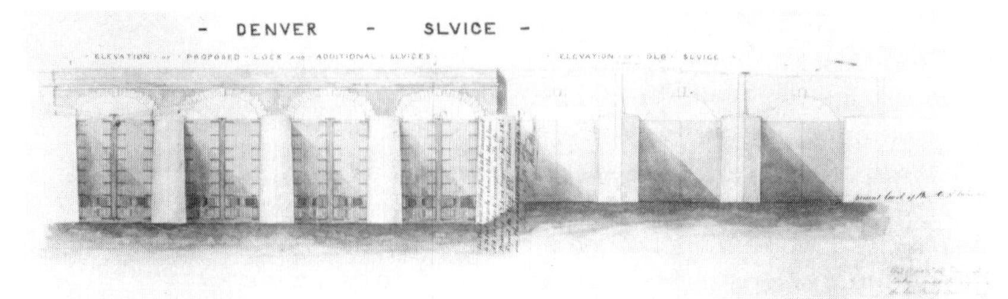

Denver Sluice, original drawing by John Rennie

19. Bedford Rivers, Old and New

HEW 820

TL 38 74 to TF 58 01

The beginning of the 17th century saw the start of a concerted effort to drain the Fens, a project indelibly linked with the name of Cornelius Vermuyden. In 1621 Vermuyden, a Dutchman, then only 26 years old but widely renowned for his land drainage works in the Low Countries, was invited to England to devise a scheme for the drainage of the Essex Marshes. Charles I had a special interest in draining the Fens and in 1630 appointed Francis, 4th Earl of Bedford, to lead a team of 'Gentlemen Adventurers' in developing drainage schemes. The Earl seized the opportunity to engage the newly knighted Sir Cornelius Vermuyden who designed and supervised the making of a new straight channel from Earith to Salters Lode on the tidal reach of the River Ouse. This cut off the great loop of the river through Ely and shortened by 10 miles the distance to The Wash. The Old Bedford River, as it became known, was the first of the great cuts across the Fens, begun in 1631 and completed in 1637, despite relentless opposition from the Fenmen. The new channel, 70 ft wide and 21 miles long had a sluice at the upper end to regulate the amount of water diverted from the river and, at the outfall end, a sluice to inhibit the inflow of tidal water from the lower reaches.

The ambitious range of drainage schemes was overtaken in 1640 by the outbreak of the Civil War and the drainage works became vulnerable both to damage due to the hostilities and the Fenmen's distrust of the new cuts.

When Cromwell took control of the Government he gave the task of further fen drainage to the recently acceded William, 5th Earl of Bedford. Vermuyden was again called upon to exercise his skills in the science of land drainage. During the 1650s a vast network of cuts, drains and sluices was developed. A major part of this work was the construction in the first half of the decade of the New Bedford River (or Hundred Foot River), which was cut ½ mile east of, and parallel to, the Old Bedford River. A flood storage reservoir of 5600 acres was created between the two rivers which became known as the Washes. Vermuyden constructed the first Denver sluice to control the outflow of the new cut into the River

Great Ouse and to exclude tidal water from the South Level rivers. The story of Denver Sluice (HEW 821) is related separately in this section.

Further reading

HARRIS E. *Vermuyden and the Fens*. 1953.

DARBY H. C. *The draining of the Fens*. CUP, 1940.

20. Ely Cathedral, Octagon Lantern

HEW 747

TL 541 802

The timber octagon surmounting the octagon tower, built in the 14th century, was one of the greatest feats of structural engineering in the Middle Ages.

The great cathedral was built by the Normans between 1080 and 1189 on the site of the earlier monastery. The tower over the main crossing in the cathedral collapsed in 1322, for reasons which are no longer apparent. This catastrophe presented the Sacrist, Alan de Walsingham, with the opportunity to build a new tower supported on eight massive columns in place of the heavy stone tower. The lofty arches, springing from the columns to span the nave and transept, carry the octagonal shell which forms the base for the lantern. This work was completed in 1328.

The timber lantern rising above this tower was designed by William Hurley, the King's carpenter. The main frame of the lantern consists of eight vertical oak posts, 63 ft high and 20 in. x 32 in. in section, each weighing about 10 tons. These had to be raised almost 100 ft above the floor to be positioned on to the octagonal sill above the arches. The lantern forms a regular octagon in plan with sides about 20 ft long. The great timber posts are braced by a complex framework of timber struts and curved rakers, and the upper compression ring is vaulted in timber.

Completed in 1334, the total weight of the lantern is approximately 400 tons.

In the 18th century James Essex undertook major repairs to the lantern, replacing some of the raker ribs which had decayed and adding extra members to reinforce the supporting framework. He also removed the external flying buttresses of the lantern. The buttresses were reinstated 100 years later by Sir George Gilbert Scott during renovation work in the course of which he restored the original Gothic outline of the lantern.

HEW 219

TL 517 730

Further reading

WADE E. C. and HEYMAN J. *Proc.Instn Civ.Engrs*, Part 1, 1985, 78, Dec., 1421–1436.

21. Stretham Pumping Station

A series of very wet years in the 1720s led to severe financial embarrassment for the Bedford Level Corporation, which was therefore unable to maintain its drainage works adequately. This led to demands from the local areas for direct control. An Act of Parliament in 1727 gave one such Level a statutory right to drain its land and impose a land tax to finance its works. Another small area gained its Act in 1738; the Waterbeach Level was the third in 1741 and many more followed.

It was under these Acts that many of the fenland windmills were erected and maintained, but wind power was not always available when needed. Agricultural improvements after the Napoleonic Wars led to a need for more reliable drainage and money was provided for the installation of steam engines. The first was installed about 1817 and in 1831 the Waterbeach Level provided one at Stretham. Built by Joseph Glynn, agent of the Butterley Co., it is now the oldest surviving in the Fens.

At first there were two boilers and in 1847 the boiler house was extended to accommodate a third. These were replaced in 1871 and 1878 by a Lancashire boiler. In 1888 the steam operating pressure was increased and new control gear installed. Inside the engine house, the engine worked for more than eighty years, until in 1909 the valve gear was updated to a design by the engine's own superintendent. Outside, the wheel is a scoopwheel, well suited to raising large quantities of water through a relatively short distance. Over the years, the ground shrank because of regular drainage and the width and diameter of the wheel was altered to suit; in 1896 it was replaced completely.

In 1925 the duties of the steam engine were taken over by a diesel, which is itself now of considerable historic interest. The old engine is cared for by the Stretham Engine Trust, who welcome visitors to this interesting relic of fen drainage history.

Further reading

HILLS R. L. *Machines, mills and uncountable costly necessities.* 1967.

22. St Ives Bridge

HEW 62

TL 312 711

The first bridge over the River Great Ouse at St Ives was a timber structure, built by Ramsey Abbey which had founded the priory of St Ivo about the year 1000 and had obtained a royal charter in 1110 to hold an annual fair there. The bridge required constant repair, so it was rebuilt in 1415–26, with six arches of limestone from

Barnack, 30 miles away on the River Welland. In 1645, during the Civil War, the Parliamentarians broke down either one or two of the southern arches and installed a drawbridge; these arches were replaced in 1716.

The four northern arches of the bridge, which date from 1426, are low Gothic spans of 13 ft to 30 ft. They have five ribs, missing in some places, supporting the arch rings which are covered by weathered drip-moulds. The two southern arches are segmental, with spans of 16 and 21 ft. The bridge has not been widened and provides a roadway of 12 ft 6 in. between the parapets.

The chief glory of St Ives Bridge is the chapel on its

St Ives Bridge

central pier. One of only three remaining in Britain, (the others are at Rotherham and Wakefield), it was dedicated to St Leger in 1426. Deconsecrated at the time of the dissolution of the monasteries in 1539, it was then used as a private house. Two extra storeys were added in 1736 but in 1930, when the whole structure had become unsafe, they were removed and the rest of the fabric was repaired and restored.

A bypass of the town was opened in 1980. Now, only traffic leaving the town is permitted on the bridge.

Further reading

BURN-MURDOCH R. *St Ives bridge and chapel*. The Friends of the Norris Museum, 1988.

Houghton Mill

23. Houghton Mill

HEW 1944

TL 281 719

A mill existed here in 974 when it was purchased as part of the endowment of Ramsey Abbey, 7 miles away to the north. One hundred years later it was one of the twenty-three mills in Huntingdonshire powered by the River Great Ouse or its tributaries. The Abbey's tenant farmers were obliged to have their corn ground at Houghton, which provided a substantial income to the Abbey. In the middle ages there were the usual conflicts between the need to impound an adequate supply of water to drive the mill and the need to allow clear flow in times of flood, which led the villagers to throw down the mill dams by force. The Abbot sought to restrain them by law, but judgment was given against him in 1515.

The mill dates from the 17th century, but was much altered in each of the succeeding three centuries. A very substantial building on an island in the river, it has five storeys of brick and tarred weatherboarding. There were three water-wheels, one on the north side driving three pairs of stones and two on the south driving seven more pairs. Much of the ancillary equipment, such as hoists , was also powered by the water-wheels.

The mill continued to operate until 1930, in the teeth of competition from imported grain and more modern technology. The water-wheels were then removed and replaced by sluices for the better regulation of the river. Belonging to the National Trust since 1939, and having been leased for use as a youth hostel for forty years, the mill is now open to the public. There is a wealth of machinery inside and one pair of wheels has been converted to electric power for occasional milling days.

24. Huntingdon Bridge

HEW 64

TL 242 715

This bridge, which Jervoise describes as 'undoubtably the finest medieval bridge in this part of England', carries the Old North Road over the River Ouse. The roadway is just wide enough for two vehicles to pass each other, so a footbridge was built in 1965–66 on the upstream side. It affords a good platform from which to inspect the fabric, which is the product of several phases of building.

There are six arches of up to 33 ft span. The arch nearest the town is lower and narrower than the others; it was apparently built early in the 14th century and widened when the rest of the bridge was rebuilt later in the century. The second and third arches are pointed, their parapets on the west side carried by ornate trefoil brackets and the cutwaters between them massive and triangular in plan. The fourth arch is segmental and the cutwater is battered back to give semi- hexagonal refuges at road level; it is probably a rebuild following the breaking down of its predecessor during the Civil War and its temporary replacement by a drawbridge. The fifth and sixth spans have arch rings and cutwaters similar to the fourth, though one is pointed and the other segmental. It is generally considered that the second and third arches were built by one authority and the fifth and sixth by another.

Although the Ouse was comparatively well supplied with bridges in medieval times, there was only a ferry at Tempsford where the modern A1 road crosses it. Huntingdon Bridge is a reminder of the former importance of the Old North Road, where, a few miles to the south, the first turnpikes in England were erected.

Further reading

JERVOISE E. *Ancient bridges of mid and eastern England*. The Architectural Press, London, 1932. 98–102.

25. Chinese Bridge, Godmanchester

HEW 1943

TL 243 706

This picturesque wooden bridge leads west from the centre of the old town of Godmanchester over a branch of the River Ouse. It was built originally in the year 1827 to the design of James Gallier and has a 60 ft span. Two segmental ribs are formed, each of six lengths of greenheart timber, butted and strapped together, and the deck is supported by diminishing saltire bracing in the spandrels.

Timber is not a very durable engineering material, being prone to decay particularly at the joints. A plaque by the bridge records that it has been rebuilt twice, in 1869 and 1960. Even so, it has been necessary to renew the handrails twice since the last rebuilding, in 1979 and 1993.

The survival of this bridge, which makes a significant contribution to an already attractive landscape, is a tribute to the authorities who have maintained it.

Chinese Bridge, Godmanchester

26. Bourn Windmill

HEW 690

TL 312 580

The first windmills in Britain were of the post mill type and those at Bourn and its neighbour Great Gransden (TL 277 554) are probably the oldest remaining, although the mill at Pitstone in Buckinghamshire (HEW 1768, section 6) is the earliest dated mill. Built almost entirely of wood, there has inevitably been a continual process of repair and renewal. Nevertheless the form of construction has remained unaltered and the mills today retain a medieval style.

At Bourn, a deed of 1653 refers to the mill as having changed hands in 1636. Successive repairs or rebuilding are evidenced by carved dates, 1742, 1758 and 1874. The buck, or body, of the mill is very small, despite having been extended to the rear at some time. Unusually it is carried by two vertical members instead of the usual

Bourn Windmill

horizontal side girts at each end of the crowntree. Two of the four sails have been replaced by shuttered sails.

The mill continued to work until 1925 when its sails were broken by a gale. It was taken over for preservation in 1931 and strengthening in modern materials has been carried out without detracting from the general appearance. Damage in the storms of 1976 has been repaired. The mill is open to the public about once a month.

Further reading

STEVENS R. *Cambridgeshire windmills and water mills.* Cambridgeshire Wind and Water-mill Society, 1985, 11.

27. Mathematical Bridge, Cambridge

HEW 469

TL 446 581

A bridge on this site in the 16th century was one of two which gave access from Queens' College to the fields on the west side of the River Cam. The other has disappeared and this one was rebuilt in 1700 and again in 1750.

The design on this latter occasion was provided by William Etheridge, who had been the foreman carpenter at Westminster Bridge, originally to have been a timber

Mathematical Bridge

bridge on stone piers. His design for Queens' was a much smaller version, in which the soffit of the bridge has the appearance of an arch, formed of parts of long straight timbers which continue into the parapets and are so jointed as to form a truss. The span is 40 ft.

Originally built between July 1749 and September 1750, decay has twice required the building of a replica structure, the last time in 1904. The main connections are bolted, as they always have been.

James Essex, the contractor in 1750, built a similar bridge (HEW 659) in Garret Hostel Lane, which he called a mathematical bridge and the name has stuck for this type of bridge.

Further reading

WILLIS and CLARK. *The architectural history of the University of Cambridge and of the colleges of Cambridge and Eton.*, **2**, CUP, 1886, reprinted 1988, 55,56.

GRAY J. H. *The Queens' College of St Margaret and St Bernard in the University of Cambridge*. CUP, 2nd Edn., 1926.

28. Clare Bridge, Cambridge

HEW 227

TL 446 584

This, the oldest bridge in Cambridge, was constructed in 1639–40. The college had just embarked on a scheme to replace entirely the 14th century buildings of which it then consisted and access over the river for materials was most desirable. One of the people working on the new buildings was Thomas Grumbold, a member of a leading family of stonemasons from the limestone district of Northamptonshire. On 18 January 1639 he was paid three shillings for a drawing of the new bridge. It is not clear now whether he was the architect or merely a draughtsman but it seems reasonable to credit him with the design. The contractor was Richard Chamberlayne and payments totalling £284.13s.8d for the bridge have been traced in the college accounts.

The bridge is one of the earliest in England of classical design, with a fine balustrade surmounted by large stone balls. It is built of ashlar limestone from Ketton, in Rutland, with some Barnack stone from an earlier construction. The three arches are of about 21 ft span and the whole structure is 75 ft long, providing a width between the parapets of 14 ft.

Over the years the bridge, and particularly the western arch, has deformed. A survey in 1972 revealed evidence of structural deficiences but, with the repairs which were

Clare Bridge

carried out at that time, it is expected that the bridge will remain adequate for the moderate loading it is required to carry.

Further reading

WILLIS and CLARK. *The architectural history of the University of Cambridge and the colleges of Cambridge and Eton.* **1**, CUP, 1886, reprinted 1988, 96.

HEYMAN J. and THRELFALL B. D. Two masonry bridges: 1 Clare College Bridge. *Proc. Instn Civ. Engrs*, 1972, **52**, Nov, 305–318.

29. Garret Hostel Bridge, Cambridge

HEW 659

TL 446 585

Until the 19th century, Garret Hostel Bridge was one of only three public bridges over the river in Cambridge. It was sited midway between Magdalene Bridge (HEW 1340) which led to Huntingdon and the Small Bridge on the Bedford road. A footbridge, it was of timber when it was recorded in 1573 and remained so through regular repairs and rebuildings. In 1769 James Essex, who had previously built the wooden bridge at Queens' College (HEW 469), designed and built a similar structure here, which he called a mathematical bridge.

After yet further rebuilding in 1814 and 1821, it was decided to have a structure of more durable material as a result of which William Chadwell Mylne designed and the Butterley Iron Company built a cast-iron bridge of 60 ft span in 1837. It lasted until 1960, but settlement caused it to fracture and it was replaced by a portal frame of 85 ft span on piled foundations. An early example of post tensioned concrete, and only 1 ft 9 in. deep at the crown, it affords easy passage for those who punt below.

Further reading

HENRY D. and JEROME J. A. *Modern British bridges.* CR Books Ltd., London, 1965, 52–53.

30. Trinity College Bridge, Cambridge

HEW 841

TL 446 586

This elegant, three-arched bridge was erected in 1763–65. It was designed by James Essex, who 13 years earlier had been working as the master carpenter on the Mathematical Bridge at Queens' College (HEW 469). One of the earliest to acquire a thorough knowledge of Gothic archi-

Trinity College Bridge

tecture, his design here is classical, with an early use of semi-elliptical arches. He was paid the considerable sum of £50 for it. The cost of the bridge itself was £1500, which was met out of a legacy from Dr Hooper, whose arms are shown in one of the shields on the bridge. The fine ashlar stonework is the product of quarries at Ketton in Rutland and Portland in Dorset. The arches span from 18 ft to 20 ft.

Further reading

WILLIS and CLARK. *The architectural history of the University of Cambridge and the colleges of Cambridge and Eton..* **2**, CUP, 1886, reprint 1988, 636–638.

31. St John's College Bridges, Cambridge

HEW 1941

TL 446 588

In 1696, work started on a replacement for the wooden bridge over the River Cam at St John's College but was suspended two years later. Building resumed on 8 May 1709 and the bridge was open by Christmas 1712. It was built by Robert Grumbold, the most prominent member of the Northamptonshire family of stonemasons which had earlier been responsible for Clare Bridge (HEW 227). Wren and Hawksmoor were consulted during the first phase of the works, particularly about the siting of the bridge, but the design is probably Grumbold's.

Built of stone from Weldon in Northamptonshire, the central semi-elliptical arch of 20 ft span has one semicircular arch on each side. The lines of the arch stones are continued radially through the spandrels, and the parapets have elegant balusters and carved panels.

When, in the 1820s, the College wished to build a new court for its undergraduates on the far bank of the river, a requirement was a bridge which would prevent access to or from the buildings after the gates were locked each night. Henry Hutchinson produced a design for a single arch of 40 ft span, supporting an enclosed footpath. It was built in 1831 and is sometimes known as the Bridge of Sighs from a slight resemblance to the well- known structure in Venice.

Further reading

WILLIS and CLARK. *The architectural history of the University of Cambridge and of the colleges of Cambridge and Eton*, **2**, CUP, 1886, 276–279.

St John's College Bridge

32. Magdalene Bridge, Cambridge

HEW 1340

TL 447 589

The bridge now known as Magdalene Bridge was formerly the Great Bridge, the original bridge over the River Cam, which gave the town its name. There was a succession of timber bridges here from Saxon times, until James Essex built the first stone structure in 1754. The cost of £1609 was financed mainly by public subscription but the County was obliged to find the money for repairs which had become necessary as early as 1788.

By 1823 the County had decided to rebuild and a Norwich architect, Arthur Browne, won the contract to design and build a cast-iron replacement. Benjamin Bevan, the canal engineer who had built the cast-iron aqueduct at Wolverton (HEW 44, section 7) twelve years previously, was engaged to ensure that the new bridge would be designed and built properly so that the County would be justified in accepting responsibility for its future maintenance.

Although by 1823 there was a substantial stock of cast-iron bridges of open spandrel design, Browne's design was archaic, a throwback to the 1797 Cound Arbour Bridge in Shropshire (HEW 423). Eight cast-iron ribs, three-centred in elevation and I- section in profile, span 44 ft between stone abutments. Cast-iron plates span between the bottom flanges of the ribs, and the spandrel walls are solid plates. The trough so formed was filled with earth on which the roadway rested directly. The spandrels were decorated with elegant Gothic tracery. The main ribs were cast in halves in Derbyshire and the secondary ironwork by Finch, a local company. The work was completed in 1824.

Over the years the large dead weight of the bridge forced the abutments apart and, being effectively a three-pin arch, the crown sagged to an unacceptable degree. After proposals to replace the iron superstructure had been rejected by local opinion, traffic was diverted in 1982 to a temporary bridge alongside. In a tricky operation in a very confined working area, the whole of the infill was removed and a steel portal frame built inside the cast-iron envelope. The façade of Browne's bridge remains to grace the area.

Further reading

MORGAN D. K. and HEATHORN T. J. A study of the design and construction and a structural analysis of Magdalene Bridge, Cambridge. *The Structural Engineer*, **59A**, 8, Aug. 1981, 255–262.

DADSON J. Cambridge builds a new bridge within a bridge. *New Civil Engineer*, 8 April 1982, 28.

33. Wicken Fen Windpump

HEW 689

TL 563 707

Of the many hundreds of pumping mills used to regulate the water levels in the fenlands only the small wind pump at Wicken Fen has survived. Erected as an iron framed mill in 1908 by Hunts of Soham, and clad in weatherboarding in 1910, this four-sided smock mill has four common sails with a total reach of 30 ft. The 13 ft diameter scoop wheel provides a maximum lift of 4 ft. The mill was operated until 1938 when it became obsolete and was left to decay.

The mill originally stood in Norman's Dyke, to serve the old turf diggings on Adventurers Fen. In 1955 it was decided to move the mill about a mile away to the National Trust nature reserve at Wicken Fen where it could be used to raise the water level in the sedge fen, a reversal of its previous drainage role. With considerable ingenuity and care the original iron frame was dismantled and re-erected at the new site, but the woodwork was completely replaced. New sails were fitted and the renovation of the mill was completed in September 1956.

Further reading

WAILES R. The windmills of Cambridgeshire. *Proc. Newcomen Soc.*

BROWN R. J. *Windmills of England*. Robert Hale, London, 1976, 60 ,61.

34. Hobson's Conduit, Cambridge

HEW 1942

TL 462 542 to TL 450 585

The original purpose of the watercourse known later as Hobson's Conduit was to provide a supply of water to flush out the King's Ditch. This had been dug about 1265 to provide a defence for the town of Cambridge and had subsequently become an obnoxious open sewer.

The watercourse was completed in 1610. Water was led from natural springs at Nine Wells, about 2½ miles south of the town, partly in a natural stream and partly in an artificial channel. By 1614 lead pipes had been installed to convey drinking water to Mill Lane and to a fountain in Market Hill. In 1631 a third pipe was added to Emmanuel and Christ's Colleges.

The channel from Nine Wells, where an obelisk was

erected in 1861, was 15 ft wide at the surface and 4 ft deep. To minimise the extent of the works, the course of Vicar's Brook was used as far as possible and the fall of 8 ft takes place in the first 2 miles. The last ½ mile, in artificial embankment, is almost level and the flow here is sluggish. The channel is puddled with the local Gault clay. It can be seen between the University Botanic Gardens and Trumpington Road, where it is spanned by four small cast-iron bridges made by Hurrell, a local company, in 1850 and 1851.

Sometime soon after 1794, the open runnels along Trumpington Street were moved to the sides of the road where they are still visible today. The fountain in Market Hill was removed to its present site at the corner of Trumpington Road and Lensfield Road in 1856 and a smaller fountain with a pumped supply replaced it.

Further reading

BUSHELL W. D. *Hobson's Conduit*. CUP, 1938.

35. London to Cambridge Railway

HEW 1733

TQ 33 81 to TL 46 57

The railway from London to Cambridge, promoted by the Northern and Eastern Railway Co. (N & ER), was incorporated on 4 July 1836, the same day as the incorporation of the Eastern Counties Railway (ECR) line to Norwich via Ipswich (HEW 1662). As events turned out, the Cambridge line had been completed and extended to Norwich four years before the ECR reached that city.

The Cambridge route had been surveyed in 1825 by John Rennie and was resurveyed in 1835 by James Walker. Robert Stephenson was appointed as the engineer with George Bidder as his assistant and the contract was let to Grissell & Peto. It had been intended to locate the London terminus at Islington but the ECR persuaded the N & ER to construct the Cambridge line from Stratford with permission to use the ECR line into their London terminus, then at Shoreditch. As the ECR had adopted a track gauge of 5 ft this change of route meant that the Cambridge line had to be built to the same gauge. Construction began in 1839 but before the line had been completed the ECR took over the lease of the N & ER on 1 January 1844 and later that year converted all their tracks to the standard gauge of 4 ft 8 ½ in. The Cambridge line was eventually rerouted from Tottenham to Bethnal Green and thence into Liverpool Street Station.

The line to Cambridge follows the River Lea valley to the north of Broxbourne where it turns alongside the River Stort to Bishops Stortford and joins the River Cam from Newport to Cambridge. This close relationship with rivers involved repeated river crossings with 17 river bridges between Bishops Stortford and Cambridge. Apart from the many river crossings there were few physical obstacles along the route although a small amount of tunnelling could not be avoided. The line passes less than a mile from Audley End House and the owner, Lord Braybrooke, required the railway to be hidden from view which was achieved by the building of two tunnels, Audley End 456 yd long and Littlebury 407 yd. Ornamental portals were added to the tunnels to compliment the noble estate.

Several stations along the northern part of the line were designed by Francis Thompson, a distinguished architect whose work survives in the listed station buildings at Audley End and Great Chesterford, but his outstanding architectural contribution was the design of Cambridge station. Here, Thompson produced a handsome colonnaded *porte cochere* of 15 arches along the front of the station. The spandrels of the arches are decorated with the arms of the colleges although the University had refused to allow the railway to build its station nearer than a mile from the town lest the serenity of the place were to be disturbed by the trains. Subsequently, the town expanded to reach and envelop the station. The station has a single platform 165 ft long, which can take two full length trains, a unique feature for a large BR station until recent years.

In the second half of the 20th century, major developments along the route such as the growth of Harlow New Town and the emergence of Stansted as the third London airport have added to the importance of the line. Electrification of the route was completed in 1988.

Further reading

GORDON D. I. *Regional history of the railways of Great Britain*, **5**, *The eastern counties*. David and Charles, Newton Abbot, 1977.

1 Cley Windmill
2 Blakeney and Cley Marshes
3 Wiveton Bridge
4 Sutton Windmill
5 Potter Heigham Bridge
6 Mayton Bridge
7 St Crispin's Road Bridge, Norwich
8 St Miles Bridge, Norwich
9 Blackfriars Bridge, Norwich
10 Bishop Bridge, Norwich
11 Foundry Bridge, Norwich
12 Ipswich and Norwich Railway
13 Lakenham Viaduct
14 Trowse Swing Bridge (old), Norwich
15 Trowse Swing Bridge (new), Norwich
16 Norwich to Yarmouth and Lowestoft Railway
17 Norwich to Lowestoft Railway Swing Bridges
18 Oulton Broad Swing Bridge
19 Berney Arms Windmill
20 St Olaves Bridge
21 Herringfleet Windmill
22 Lowestoft Lighthouse
23 Garrett's Long Shop, Leiston
24 Homersfield Bridge
25 Grime's Graves
26 Town Bridge, Thetford
27 Pakenham Watermill
28 Pakenham Windmill
29 Bury St Edmunds Station Bridge
30 Moulton Packhorse Bridge

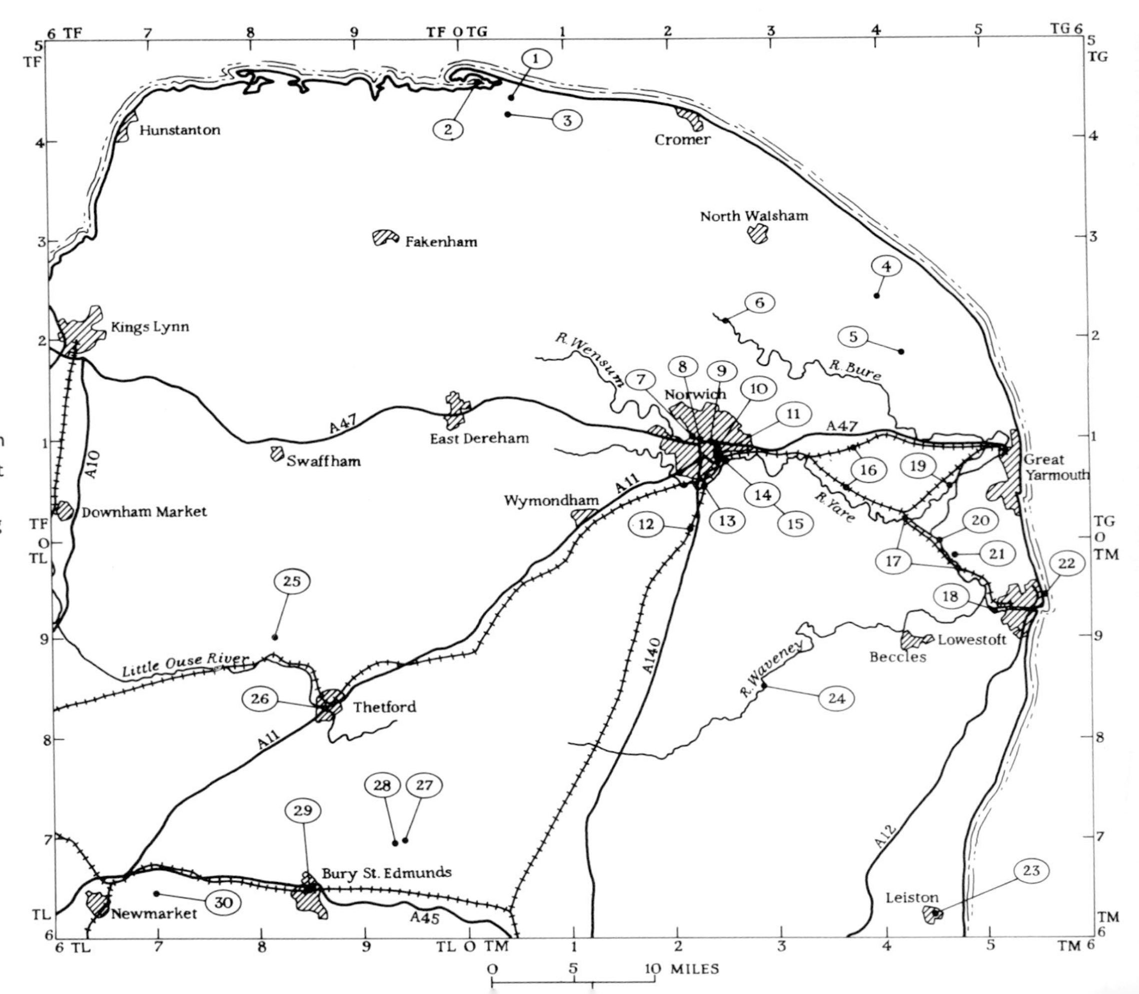

4. Norfolk and North Suffolk

The gently undulating terrain of East Anglia presents little need for major structures to carry roads and railways. The two counties can boast few large bridges although Norwich has a fine sequence of attractive bridges over the River Wensum. Particularly distinctive are the medieval Bishop Bridge and St Miles Bridge, the oldest cast-iron span in East Anglia. The lack of local stone led to a widespread use of the plentiful supply of flints for building work, supplemented by bricks. The oldest known civil engineering site in the area is Grime's Graves, near Thetford, where Neolithic men mined flints for tools and weapons some 4000 years ago.

Traversing eastwards, the open heathlands and forests of Breckland give way to rich agricultural country and then to the remarkable feature of Broadland, 200 miles of navigable water with 30 broads, irregularly shaped lakes which formed when medieval peat diggings became flooded. The Broads are now the key element of one of the great leisure areas of the country. The slow-moving, shallow rivers are generally unsuitable for water-mills and, since early times, windmills have been a feature of the landscape. There were once 850 windmills and 100 wind pumps in Norfolk alone but now only a handful remain. The crumbling towers of old mills are still scattered across the countryside but a few survivors like Berney Arms mill make a proud sight, with their sails turning at times. Suffolk has its own heritage of mills and important examples are to be found at Herringfleet and Pakenham.

Crossing the county boundary at Homersfield, a bridge of national historic importance carried the road across the River Waveney until a bypass was constructed. The earlier bridge is the oldest iron-reinforced concrete bridge in the country.

The distinctive coastline sweeps round the great Anglian headland with few breaks except in the north. Lighthouses were an early feature of this often storm-tossed coast and Lowestoft takes pride of place as the most easterly land light on the British coastline.

1. Cley Windmill

HEW 1581

TG 045 441

A notable and picturesque landmark on the North Norfolk coast is the tower windmill at Cley next the Sea. Standing on the old quay alongside the River Glaven, it

Cley Windmill

is the most distinctive feature of the village which was an important seaport during medieval times.

A disastrous fire destroyed much of the village in 1612 and new buildings, including a post mill, were then erected around the wharves at the riverside. The tower mill came much later and was first recorded as 'newly erected' in 1819. The Cley millers appear to have prospered. In the first half of the 19th century John Lee, who rented the mill, also owned seven cottages and a house in the village. He was followed in 1848 by Lawrence Randall who widened his activities over the next 25 years on a grand scale to become 'miller, corn, coal and timber merchant, cabinet maker and proprietor of the George & Dragon Inn'. The last, and perhaps best known, of the millers were Steven Barnabus Burroughs and his sons, who owned and worked the mill until 1917, by which time the silting up of the channel had resulted in the decline of trade at Cley and the mill ceased work.

The tower is some 50 ft high to the white painted conical cap with its eight bladed fantail and four sails having a total span of 68 ft. It has four storeys and tapers from 26 ft to 18 ft in diameter with an elegant gallery at first floor level. In 1921 the mill was bought by Sarah Maria Wilson and has since been used as living accommodation.

Further reading

BROWN R. J. *Windmills of England*. Robert Hale, London , 1976, 148–149.

2. Blakeney and Cley Marshes

HEW 1756

TG 02 44 to TG 04 44

A mile north of the A149 Coast Road where it passes through Blakeney and Cley next the Sea a great shingle bank sweeps in a north-easterly direction at an angle to the land, ending in Blakeney Point. Within the shelter of the shingle bank the Blakeney and Cley Channels follow their tortuous routes flanking enclosed areas of fresh water marshes. For several centuries the marshes have been protected by banks, although these defences have been overwhelmed at intervals by exceptionally high tides, notably in 1742, 1897 and 1953. On each occasion the embankments were breached and had to be rebuilt and strengthened.

The banks stand some 12 ft to 18 ft high and vary from 6 ft to 10 ft wide at the top. The core of the banks is protected by an impervious membrane and the battered slopes are turfed. The bank around Blakeney Marsh is 2 ½ miles long and the Cley Marsh is slightly shorter at 2 miles. The first embankment and drainage of the Cley and Salthouse marshes was begun by Sir John Heydon in 1522 and this work was improved and extended by the Dutchman, van Hasedunch, starting in 1630. The enclosure of the Cley Marshes was completed in 1649 by Simon Britiffe, Lord of the Manor, who built the bank on the east side of the Cley Channel.

In the Middle Ages the ports of Blakeney, Cley and Wiveton, known collectively as 'The Haven', were busy and prosperous harbours both for fishing and for trade with the Low Countries. However, the progressive enclosure of the marshes caused havoc in the channels leading to the ports. These channels had remained navigable, because of the scouring effect of the run-off from the marshes with each ebb tide, but the elimination of this discharge, resulting from the enclosures, caused silting of the channels and ultimately led to the end of sea trading.

The Cley Marshes are now noted as a bird sanctuary and the whole area is part of a designated Heritage Coast.

Further reading

STEERS J. A. *Coastline of England & Wales*. CUP, 1964, 352–354.

Wiveton Bridge, west side

3. Wiveton Bridge

HEW 1541

TG 044 427

The stone bridge over the River Glaven at Wiveton, near Blakeney, has an interesting pointed arch of a type not found elsewhere in North Norfolk. Sturdily built in dressed stone blockwork, it dates from the 14th or 15th century. A distinctive feature of the 32 ft span is the strengthening of the arch with five chamfered ribs, a form of construction also used in Toppesfield Bridge, Suffolk (HEW 837).

The earliest known reference to the bridge was in the bequest of Robert Paston of Wiveton in 1482 for 6d. to repair the chapel on the bridge, a building which no longer exists. Repairs to the bridge by the Council were first recorded in 1634 but the patches of replacement stone in the parapets are evidence of more recent repairs as are the brick pilasters at the parapet ends, probably the aftermath of vehicle contacts with the walls on this narrow hump backed bridge.

This lane was once the first inland route along the coast until the new road through Cley was opened in 1824 and would have been used extensively when Wiveton was a tidal port during the 13th to 16th centuries. Now the bridge, an Ancient Monument, stands half forgotten in the meadows. Over the centuries the structure has survived the periodic floods of the Glaven and the more serious incursions of the sea, the most significant inundation in recent times being the East Coast floods of 1953.

Further reading

JERVOISE E. *The ancient bridges of mid & eastern England*. The Architectural Press, 1932. 115.

4. Sutton Windmill

HEW 1533

TG 396 239

The tallest tower mill in the country, 79 ft high, Sutton windmill stands in the flat land east of the village of Sutton and was built in 1789 by a millwright, named England, of nearby Ludham. The brick tower has nine storeys with four pairs of French burr millstones located on the fifth floor. The round tower, 32 ft 6 in. wide at the base tapering to 16 ft wide at the top, supports the cap with four sails and a ten bladed fantail. The cedar clad cap is boat shaped, a distinctive feature of many Norfolk mills.

The mill was severely damaged by fire in 1857 but was re-equipped and fitted with patent sails in 1860. It remained in use until 1940 when it was struck by lightning and the sails burnt. The present owner, Chris Nunn, has carefully renovated the mill including renewing the massive 66 ft long sail stocks in Columbian pine. The Listed Grade II building and its exhibition of rural industries are open to public visits during the summer months.

5. Potter Heigham Bridge

HEW 1543

TG 420 185

The main road between Great Yarmouth and Cromer crossed the River Thurne over a 14th century stone-built bridge of three arches, now happily relieved of heavy traffic with the diversion of the A149 North Walsham to Great Yarmouth road over a modern bridge a short distance upstream. The old bridge was for centuries the only crossing of the Thurne between the combined outflow from Hickling Broad and Heigham Sound to the east and Acle Bridge to the south-west, a river distance of some 7 miles.

The central segmental arch of 21 ft span has a clearance to the crown of only 7 ft and requires cautious navigation by the busy launch traffic on the river. Jervoise

Potter Heigham Bridge

considered this arch was of a later date. The two sharply pointed side arches, 8 ft 3 in. wide and partly submerged, have chamfered ribs in the soffits and are evidently part of the original structure. The stone spandrels are built in semi-random blocks with shallow buttresses over the piers but the old stone parapets and refuges were rebuilt in brickwork about 1920. The brick retaining walls to the raised approaches are heavily buttressed and there are curious cantilevered buttresses at the abutments.

The bridge, an Ancient Monument, bears the scars of both river and road traffic. The main arch has been scored by many passing boats and the spandrels bulge outwards from the legacy of heavy road traffic.

Further reading

JERVOISE E. *The ancient bridges of mid and eastern England.* The Architectural Press, 1932, 116.

6. Mayton Bridge

HEW 1542

TG 250 216

A mile to the south-east of Buxton the country lane to Little Hautbois crosses the River Bure over an early Tudor period bridge with raised approaches giving a sharp hump in the road. Known as Mayton Bridge, perhaps by association with the nearby moated Mayton Hall, it is a small red brick structure of two arches which are unusual in being four centred in profile. Despite its lack of symmetry with unequal spans of 12 ft 9 in. and 9 ft it is pleasantly proportioned.

A most unusual, perhaps unique, feature of the bridge is a brick shelter or wayside shrine with a low arched opening and tiled roof at each end of the upstream parapet. The bridge is an Ancient Monument and has recently been sympathetically repaired.

Further reading

JERVOISE E. *The ancient bridges of mid and eastern England.* The Architectural Press, 1932, 115.

7. St Crispin's Road Bridge, Norwich

HEW 1124

TG 226 092

The Eastern & Midlands Railway (later the Midland and Great Northern Joint Railway) opened Norwich City Station in 1882 as the terminus of their new line from Melton

Constable and built the Station Bridge over the River Wensum in the same year to carry the station approach road. City Station closed in 1969 and the bridge now serves the south carriageway of the ring road; an adjacent modern structure carries the north carriageway over the river.

St Crispin's Bridge, as it is now named, has a flat-arched wrought iron span of 55 ft with latticed spandrels and a width of 26 ft. The cast-iron open patterned parapet balustrade adds a note of distinction to the bridge. The span was fabricated by a local firm whose name is exhibited on the parapet plinth in the inscription, NORWICH BARNARD, BISHOP & BARNARDS 1882.

8. St Miles Bridge, Norwich

HEW 1125

TG 227 088

A cast-iron road bridge dated as early as 1804 is now exceptional and St Miles bridge is thought to be the oldest of this type in East Anglia. This narrow crossing of the River Wensum once carried busy traffic of the adjacent brewery (now defunct) and of the traders in Coslany Street but much of the area has since been converted to

St Miles Bridge, Norwich

residential use and the bridge is now restricted to pedestrians and cyclists.

The arched span of 36 ft 2 in. is made up of four cast-iron ribs each assembled from five bolted sections with solid spandrels and parapets. The centre panel on the outer face of both parapets bears an embossed plaque depicting the city arms and the date of 1804 is cast marked at the centre of the arch. A spout for hoses projects from the west parapet.

The bridge was designed by James Frost, of St Faith's Lane, Norwich and it was built at a cost of £1100. His modest bridge has survived the tests of nearly two centuries including the flood in 1912 which reached parapet level. The bridge has been tastefully restored (apart from the discordant feature of the tubular handrails) and the complete width of 15 ft between the parapets has been paved in brick.

9. Blackfriars Bridge, Norwich

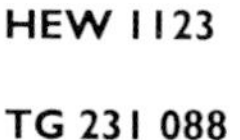

HEW 1123

TG 231 088

Blackfriars Bridge takes its name from the Dominican monks who established an extensive monastery here in the 13th century. They evidently built a timber bridge

Below: Blackfriars Bridge, Norwich

across the River Wensum which was rebuilt in the time of Edward IV and in turn this was replaced in 1589 by a three-span stone structure.

The present bridge, constructed in 1784, was designed by Sir John Soane, the architect noted for his design of the Bank of England. The Wensum bridge was built by John de Carle, a stonemason of Norwich. Soane's bridge has a single masonry segmental arch of 44 ft 6 in. span with voussoir facing to the arch and masonry spandrels. The section of St George's Street across the bridge is very narrow, being only 10 ft wide over the span. The footpaths are supported on cantilevered iron ribs which also carry the highly ornate cast-iron parapet balustrades. The west parapet is pierced by a hose spout, a feature found in several other river bridges within the city limits.

Further reading

JERVOISE, E. *The ancient bridges of mid and eastern England*. The Architectural Press, 1932, 120.

10. Bishop Bridge, Norwich

HEW 1776

TG 240 090

Bishop Bridge is the only remaining medieval bridge over the River Wensum in Norwich and the most historically important. In 1275 Edward I granted a patent to the Prior of Norwich, William de Kerkeby, to build a gate with a bridge across the river, replacing an earlier timber bridge.

In the late 13th century a distinguished citizen of Norwich, Richard Spynk, built the stone and brick bridge with three segmental arches strengthened with stone ribs which are infilled with brickwork. The spans vary from 15 ft to 25 ft and the stone piers carry pointed cutwaters at both ends. The gate, known as the Ethelbert Gate, was built at the west end of the bridge with a fortified gatehouse extending onto the bridge over two of the arches. The bridge and gate were in the ownership of the priory until 1393 when the city formally took possession and rough stone tablets in the parapets display the city arms.

An examination of the structure in 1790 revealed serious weaknesses caused by the weight of the gate tower and this was demolished. The roadway over the bridge is wide enough only for single file traffic with footpaths on both sides, but the carriageway may have been wider in earlier days and there are interesting refuges in the parapets, probably for pedestrians.

Further reading

JERVOISE. E. The ancient bridges of mid and eastern England. The Architectural Press, 1932, 121–122.

11. Foundry Bridge, Norwich

HEW 1777

TG 238 085

Foundry Bridge, a wrought iron single span bridge of 1886 over the River Wensum, carries the wide thoroughfare which links the railway station and the city centre. When the first bridge was built here in 1810–11 it was a rural site, apart from an old iron foundry which gave its name to the bridge. This bridge had a timber deck on stone piers and served the occasional horse-drawn cart on the byway until the arrival of the railway stimulated development in the area.

In April 1844 the first railway in East Anglia was

Foundry Bridge, Norwich

opened between Norwich and Yarmouth (HEW 1685) and the railway station was built on the east side of the river. In the same year the earlier bridge was replaced by a cast-iron structure providing a more appropriate access between the terminus and the city, the cost being shared between the city and the Norfolk Railway Co.

The present bridge was constructed in 1886 at a cost of £12 000 and a contribution was made by the Great Eastern Railway Co. The wrought iron span of 55 ft comprises four main girders 4 ft 2 in. deep and has a generous width of 50 ft between the ornate parapets. The main girders were fabricated in the station yard and launched sideways into position from the old bridge.

12. Ipswich and Norwich Railway

HEW 1662 (PART 2)

TM 157 438 to 230 077

As the railway from London to Ipswich neared completion in June 1846, pressure mounted for further lines into the heart of East Anglia. Locke and Bruff, with Brassey as contractor, built a railway from Ipswich to Bury St Edmunds in a remarkably short time, opening the service at Christmas 1846. The same team then began the final leg of the Eastern Union railway to Norwich in February 1847, making a junction with the Bury line north of Stowmarket.

A notable structure on the Norwich line is the brick viaduct at Lakenham (HEW 1444) across the River Yare and the Cambridge line. At Stowmarket the station buildings were designed in an attractive Jacobean style by the Ipswich architect, Frederick Barnes.

The service to Norwich finally opened on 12 December 1849, some 13 years after the project for the railway from London had begun. The Norwich terminus was Victoria Station and it was not until the 20th century that the Ipswich line was linked to the Cambridge route to divert trains into Norwich Thorpe station, crossing the swing bridge over the River Wensum at Trowse (HEW 743).

Further reading

GORDON D. I. *Regional history of the railways of Great Britain, The Eastern Counties*. David and Charles, Newton Abbot, Chps. 2 & 4.

13. Lakenham Viaduct

HEW 1444

TG 228 053

On the southern edge of Norwich a fine brick arched viaduct carries the main line railway from Ipswich across the River Yare and over the railway from Cambridge. The viaduct was built between 1847 and 1849 as part of the works for the railway from Haughley Junction near Stowmarket. The line was then known as the Norwich Extension which completed the rail route from London, Liverpool Street, via Ipswich (HEW 1662). The engineer for the extension was Joseph Locke, assisted by Peter Bruff and the work was carried out by the contractor Thomas Brassey.

Locke's viaduct is 312 ft long with six segmental arches varying from 42 ft to 45 ft in span on a 25° skew and rises 40 ft clear above the river, the largest bridge on the Norwich Extension line. The plain but well proportioned structure was built in red brickwork except for the 'Suffolk white' brick facing to the six rings of each arch, a feature to be found on some other bridges built for the Eastern Union railway.

14. Trowse Swing Bridge (dismantled), Norwich

HEW 743

TG 245 076

The Eastern Counties Railway (ECR) from Bishops Stortford via Cambridge and Ely, engineered by Robert Stephenson and George Bidder, was opened as a through route to Norwich in 1845. To reach Norwich Thorpe Station, which had been built to serve the 1844 line to Yarmouth, the ECR had to cross the River Wensum with restricted headroom (about 9 ft) ½ mile from Thorpe Station. Crossing this navigable waterway was accomplished by constructing a swing bridge, designed by Bidder and built by Grissell & Peto. This was one of the first railway swing bridges. The single track bridge was supported on a central pier founded on cast-iron piles and the deck turned on a roller path within a cast-iron headplate on the pier.

A replacement bridge to carry two tracks was installed in 1905 incorporating the centre pier and headplate of the 1845 bridge.

It had 120 ft long main girders with central portals 23 ft long by 19 ft high connected by 3 in. diameter steel tie-rods to the girder ends. Four additional steel cylinders were sunk outside the old centre pier to support a new roller path and the bridge rotated on two sets of cast-iron rollers and a 7 in. centre pivot, presenting a river opening 43 ft wide. The contractor was Andrew Handyside of Derby. This second bridge was replaced in 1987. (HEW 1597).

Further reading

BIDDER G. P. Cast-iron swing bridge over the River Wensum at Norwich. *Min. Proc.Instn Civ.Engrs. 1846,* **V**, 434–438.

15. Trowse Swing Bridge (1987), Norwich

HEW 1597

TG 245 076

The new railway swing bridge over the River Wensum was constructed alongside the 1905 bridge (HEW 743) during the period October 1984 to February 1987 to meet requirements for the electrification of the route. The track format reverted to a single line as on the original 1845 bridge but in the new design the eccentric swing span

Trowse Swing Bridge

pivot is located on the south bank instead of a central support on the river pier. The superstructure is a welded steel through-type bridge with steel cross girders and rail bearers carrying greenheart timber waybeams.

The swing span weighing 360 tonnes has a 20 m nose balanced by a 6 m tail with a 200 t kentledge and rotates on a Roballo sluing ring with ball races supported on a steel foundation ring. The span is simply supported in the closed position and is raised 300 mm on hydraulic jacks before rotation. Portal frames carry the 25 kV overhead electric contacts which are independent of the contiguous overhead installation. This was the first swing bridge with overhead electrification on British Railways.

The bridge was designed by Rendel Palmer & Tritton and the construction was carried out by May Gurney & Co. After the new bridge had been brought into service on 15 February 1987 the old swing bridge was dismantled.

Further reading

LEWIS W. M. and CLARK P. J. Railway electrification: Anglia civil engineering projects. *Proc. Instn Civ. Engrs*. 1987, I, 261–268.

16. Norwich to Yarmouth and Lowestoft Railway

HEW 1685

TG 239 083 to TG 519 081

The line from Norwich to Great Yarmouth was the first railway in Norfolk. George and Robert Stephenson readily supported the proposal launched by the Yarmouth & Norwich Railway Co., which was later renamed The Norfolk Railway. George Stephenson was elected Chairman of the Directors and the project received Royal Assent in June 1842. Robert Stephenson was appointed as the engineer, assisted by George Bidder and work started in April 1843 with Grissell & Peto as the contractors. Fifteen hundred men were employed on the work.

At Norwich the River Yare was diverted for ½ mile to avoid the need for a bridge and the line followed the river via Reedham and then turned north-east to skirt Breydon Water, before curving sharply into Yarmouth, a distance of 20½ miles. The railway was completed at a cost of £200 000 in the remarkably short time of twelve months, six months less than estimated. A special train travelled along the new line on 12 April 1844. This was the first railway in the country to be equipped with electric telegraph, the Cooke & Wheatstone system, which provided an early form of block signalling. An unusual station on the section across the Reedham marshes is a request halt which serves only Berney Arms windmill (HEW 686).

Samuel Morton Peto, later MP for Norwich, was anxious to promote the development of Lowestoft as a major port and helped to sponsor the rail connection to the town. The 11 mile long connection from Reedham was built by George Bidder who is credited with the design for the first railway swing bridge. Two of these structures were required on the Lowestoft line, one at Somerleyton (HEW 745) and one at Reedham (HEW 746).

The Norfolk Railway was merged with the Great Eastern Railway in 1862. By the 1870s the rising tide of holiday visitors to Yarmouth led to the doubling of the line between Brundall and Reedham and this was followed in 1883 by a new line on a shorter route from Brundall via Acle to Breydon Junction. The first terminus at Norwich was closed when the present Thorpe Station was built in 1886. This imposing building was designed by Wilson and Ashbee in a free Renaissance style.

Further reading

GORDON D. I. *Regional history of railways of Great Britain, Volume 5*. David and Charles, Newton Abbot, Chapter 8.

17. Norwich to Lowestoft Railway swing bridges: Reedham Swing Bridge and Somerleyton Swing Bridge

Reedhan Swing Bridge:
HEW 746
TG 422 017

Somerleyton Swing Bridge:
HEW 745
TM 476 967

The railway between Norwich and Lowestoft is carried over the River Yare and Waveney at Reedham and Somerleyton by two similar swing bridges, built in 1905 to replace single track bridges.

In the fixed position for rail traffic, the live load is carried by three wrought iron girders, 139 ft long, bearing on the central pivot pier and the two end piers. When a bridge is opened for river traffic, the cantilevered load is carried by two truss girders, 19 ft deep at the centre and 7 ft deep at the ends.

Somerleyton Swing Bridge

Each span rotates on 16 in. diameter cast steel wheels and a 10 in. diameter central pivot, resting on a brick central pier of 27 ft 6 in. diameter, supported by timber piles. The approach spans are supported by wrought iron girders.

18. Oulton Broad Swing Bridge

HEW 744

TM 522 927

Oulton Broad Railway Swing Bridge is situated at the west end of Lake Lothing, Lowestoft, a tidal water open to the sea. The bridge carries the double track of the Ipswich to Lowestoft line over the narrow channel between Oulton Broad and Lake Lothing. It was built by the Great Eastern Railway in 1907 as a replacement for a single track bridge. The design of the swing span is identical to the 1905 Trowse Bridge at Norwich (HEW 743).

The substructure has brick piers on timber piles with a pivot pier of 27 ft 6 in. diameter. In addition to the swing span there are two side spans of 45 ft and one of 30 ft.

Berney Arms windmill and scoopwheel from the west

19. Berney Arms Windmill

HEW 686

TG 465 049

The flat marshlands to the west of Great Yarmouth are dotted with the crumbling stumps of derelict windmills but the lofty tower mill at Berney Arms, now carefully restored, is a lone survivor. It stands in an isolated position near the head of Breydon Water on the north bank of the River Yare.

The mill was built by Stolworthys, the millwrights of Yarmouth, sometime before 1870 and was initially used to grind cement clinker for a nearby cement works which closed in 1880. The machinery was then converted to drive a water-wheel which lifted water from a marsh cut to the higher level of the river and the mill continued to fulfill a drainage role until 1948 when it was superseded by motor driven pumps.

The 70 ft high tapered brick tower has a boat shaped cap typical of Norfolk mills, with a fantail and four patent sails. An iron gallery circles the tower at half height and another gallery surrounds the cap. At the base of the tower a horizontal shaft drives the external scoop wheel which is unusually large, having a diameter of 24 ft. The highest drainage mill in the Broadlands area, it is classified as an Ancient Monument and is cared for by the Department of the Environment who open it to public visits in the summer months. The mill is unique in having its own railway station, a request halt on the single line between Reedham and Yarmouth, but there is no road access. Few visitors arrive by rail and most come from boats attracted by this prominent landmark and the novelty of the turning sails.

Further reading

BROWN R. J. *Windmills of England. Robert Hale.* London, 1976, 144.

20. St Olaves Bridge

HEW 192

TM 457 994

The A143 road between Great Yarmouth and Beccles crosses the River Waveney at St Olaves by a cast-iron tied-arch bowstring bridge, a classic example of its type and unique in East Anglia. The bridge has a clear span of 80 ft and the bowstring girders, at 24 ft centres, rise to a height of 15 ft above the deck. The main girders are

St Olaves Bridge

formed from sets of four cast-iron T section ribs which are braced by cast diaphragms bolted to the ribs at 5 ft intervals making a lattice assembly girder 2 ft 6 in. high and 2 ft 3 in. wide. The 45 ft long girder segments are bolted at a friction joint in the crown of the arch.

The deck was originally supported on cast-iron cross girders suspended from the main girders by 1¾ inch diameter wrought iron hangers but the deck girders were replaced by steel joists in 1959. The main ribs are tied by flat link wrought iron chains. The bridge was built in 1847, by the contractor George Edwards of Carlton Colville, Lowestoft.

21. Herringfleet Windmill

HEW 688

TM 466 976

Herringfleet Windmill, a little black smock mill and the last surviving example of this type, stands alone on the east bank of the River Waveney across open meadows ½ mile from the B 1074 road between Somerleyton and St Olaves. It was built about 1820 by Robert Barnes, a Yarmouth millwright.

The small octagonal tower has a wooden frame with tarred weather-boarded walls. A boat shaped cap supports four sails, which were cloth covered in working order, the sails turning anti-clockwise. On the other side of the cap is a braced tailpole used to steer the cap and the sails into the wind, aided by a hand winch at the lower end of the tail pole.

Inside the mill there are three floors, the ground floor being the spartan accommodation for the marshman who would sometimes have to spend all night in the mill if the weather demanded constant attention to the sails. From the brakewheel on the iron windshaft in the cap a stout vertical wooden shaft transmits the drive to the ground level where a horizontal iron shaft turns an outside scoopwheel housed in a semicircular casing. The 16 ft diameter scoopwheel with 9 in. wide wooden paddles lifted water from a drainage channel 10 ft to discharge into the river.

The Somerleyton Estate kept the mill working until 1955 when an electric pump was installed. Fortunately, Suffolk County Council decided to restore and preserve the smock mill which is kept in working order and is operated occasionally for visitors.

Further reading

FLINT B. *Suffolk windmills*. The Boydell Press, Suffolk, 1979, 90–91.

Herringfleet Windmill

22. Lowestoft Lighthouse

HEW 750

TM 551 943

Lowestoft Lighthouse has the double distinction of being the oldest established lighthouse station still in use on the coast of Great Britain and our most easterly shore light. Navigation lights have been displayed from this site for nearly 400 years except for interruptions during war years. The original double lights were erected on the foreshore by a local venturer named Bushell and were taken over by the Trinity Brethren in 1609. The primary purpose was to provide a bearing for ships using the Stanford Channel, a restless and fickle passage through the offshore sands, but the Lowestoft lights also became an important beacon for passing ships.

Soon after Samuel Pepys was appointed Master of Trinity House in 1676 the decision was taken to build a new tower lighthouse at the top of the cliffs. Pepys took a personal interest in the project and a stone plaque displaying the Trinity House coat of arms above the heraldic device of Pepys was fixed on the tower. When the building was demolished 200 years later the plaque was preserved and is a prized exhibit in the lighthouse today.

The present circular tower, 59 ft high was built by

Lowestoft Lighthouse

Suddelay & Stanford of Beverley, Yorks, at a cost of £2350 and was brought into service in 1874. Electric lighting was intended and the new tower had been built to take the extra weight of electrical equipment. However, the lantern was at first fuelled by paraffin oil followed by gas and it was not until 1936 that the light was converted to electricity. Standing 123 ft above sea-level, the light is one of the most powerful on the East coast.

Further reading

LONG N. *Lights of East Anglia.* Terence Dalton Ltd, 1983, 87–109.

23. Garrett's Long Shop, Leiston

HEW 840

TM 444 625

Now part of an industrial museum, the Long Shop at the Garrett Works in Leiston was the first machine assembly factory building designed for flow-line production in the United Kingdom. Founded in 1778 by Richard Garrett, the firm became widely known for its agricultural machinery and by the middle of the 19th century was expanding to meet the growing demand for steam powered machines. In 1851 during a visit to the USA, Richard Garrett III saw continuous line production and decided to adopt similar methods for the assembly of large ma-

Garrett's Long Shop interior

chines at his works. In the following year he began building his new workshop, completing it in 1853.

The Long Shop, known locally as 'the cathedral', is a brick building 85 ft long, 41 ft wide and 43 ft high. Inside, the most distinctive feature is an 11 ft wide timber gallery around the central area which is open to the full height of the building. The supporting frame of the gallery is made of 11 in. square timber posts with double cast-iron brackets spanning between the posts at gallery deck level and cast-iron knees holding the cross beams. Two of the double brackets bear the cast inscription 'RG 1853'. Two travelling cranes (not original) run on rails above the gallery frame. The assembly crew worked at two levels on the large traction engines, steam rollers and tractors as they progressed through the workshop until the completed machines emerged through the high doors at the end of the building.

Fortunately the Long Shop was kept in its original form when machine assembly was moved to another part of the works. Since the factory closed in 1980 the Suffolk Preservation Society has worked to create the museum and to ensure the survival of this unusual and impressive building.

Homersfield Bridge, detail of arch

24. Homersfield Bridge

HEW 836

TM 283 857

There are few remaining examples of early attempts to combine iron and concrete in bridge construction before the emergence of true reinforced concrete at the end of the 19th century but a remarkable survivor is the 50 ft span over the River Waveney at Homersfield on the Suffolk/Norfolk boundary. It is evidently our oldest hybrid reinforced concrete bridge. Built in 1870 for the owner of Flixton Hall estates, Sir Shafto Adair, the arch is reinforced with a wrought iron framework encased in concrete. This frame consists of vertical lattice girders formed from 6 in. x 4 in. angles linked at soffit level by shallow iron joists fixed 2 ft apart. The monolithic arch, segmental in profile, rises 5 ft and tapers from a depth of 6 ft 4 in. at the haunches to 2 ft 3 in. at the crown. The cast-iron balustrades are ornate with Adair monogrammed bosses at the panel centres and decorative finial caps on the posts.

Documents of the time indicate that the bridge was designed by the contractors, Messrs W & T Phillips of London, to instructions given by Henry Eyton, an Ipswich architect, who was the agent for Sir Shafto. Eyton's name is embossed on the handrail of a balustrade. The quotation for the work was £344, 'a low price compared with some of the County bridges' wrote Eyton in a letter to Sir Shafto on 17 December 1869.

The bridge once carried the busy B1062 Flixton to Wortwell road, but the road was diverted over a new bridge in 1970 and the old one is now used as a foot and cycle path. In earlier days the Flixton Estates placed a chain across the bridge on one day each year and claimed a toll of a penny per wheel.

The bridge is Listed Grade II*. After years of neglect the Flixton Estates Bridge is now recognised as the oldest concrete bridge in the UK and the Norfolk Historic Buildings Trust is sponsoring restoration and future maintenance.

Further reading

BOSTICCO M. Early concrete bridges in Britain. *Concrete*, Sept. 1970, 363.

25. Grime's Graves

HEW 472

TL 818 898

The Breckland district of south-west Norfolk has an intriguing example of primitive engineering in an area of Neolithic flint mines known as Grime's Graves (Grim's Hollows). The pock-marked slope of a shallow valley reveals evidence from slight depressions of 700–800 pits which have been radio-carbon dated from objects found in the pits as being dug during the period 2330 to 1740 BC. The true nature of these pits was revealed by excavations conducted by a determined clergyman, Canon Greenwell, during three years work in 1868–70. He correctly deduced that the pits had been dug in the Neolithic period.

The pits were cut to reach a layer of hard chalk containing a thin vein of hard flints known as 'floorstone', some 30 to 40 ft below ground level. The chalk is overlaid with beds of boulder clay and sand through which the miners dug a funnel shaped pit tapering down to the chalk level. They were then able to sink a vertical shaft through the chalk to the floorstone. The deepest of the shafts is about 40 ft with a diameter varying between 13 and 26 ft. In some shafts the hard chalk was undercut to form a bell-shaped working cavity. At floorstone level, radiating galleries were cut to extract more flints and in Pit 1 there are 27 galleries of which nine main headings are about 5 ft high. The shafts were sunk in close proximity to each other with galleries projecting into the interstices between those of adjoining pits. Extensive beds of stone chips on the surface reveal that the flints were knapped at the site to form axe heads and other cutting tools.

The engineering significance of the pits lies in the technique adopted of battering the excavation through the softer upper soils at the angle of repose and in the use of tools. The two main types of tool used were picks made from the antlers of red deer, examples of which have been found in the pits, and shovels made of wood or from animal shoulder blades. An assessment of these tools indicates that they could have been used with an efficiency level of 70% compared with modern hand tools. Something like 100 to 150 picks must have been used for each shaft and 50 000 antlers would have been used in total at Grime's Graves. One gallery which had been blocked by a roof fall was opened up to reveal antler picks lying on the working floor just as they had been left by the miners some 4000 years ago.

The total area mined was about 80 acres although less than half of that area is in the care of the Department of the Environment. A few of the pits have been cleared and visitors can see the site exhibition and go down one of the pits.

Further reading

CLARK R. *Grime's Graves*. HMSO, London, reprinted 1975.

26. Town Bridge, Thetford

HEW 1641

TL 868 831

The Town Bridge near the centre of Thetford has a finely proportioned cast-iron arch of semi-elliptical profile, with a span of 33 ft, across the Little Ouse River. The south-east face of the arch bears the date 1829 and the other side carries the emblem of a castle. The deck has six cast-iron ribs 1 ft 3 in. deep with 1½ in. thick webs cast in four segments connected by bolted joints.

The road narrows to 15 ft over the bridge, with a marked hump, and once carried the traffic on the main London to Norwich route. It is fortunate for the survival

Town Bridge, Thetford

of the bridge that the A 11 trunk road has been diverted over a new route through the town. This relief from heavy traffic and the discreet strengthening of the deck with reinforced concrete arches between the cast-iron ribs, which was carried out in 1964 by the Norfolk County Council, has greatly extended the life of the bridge. The structure is brightly painted in red, green and white and has decorative parapet railings with ornate lamp posts at the centre of the span. It is a Listed Grade II structure. The bridge was built by the firm of Brough & Smith from City Road, London.

27. Pakenham Water-mill

HEW 838

TL 937 695

There are now few working examples of the water-mills which have been a feature of some East Anglian rivers during the past 1000 years, but there is a distinguished exception at Pakenham. A mill was recorded here when the Domesday survey was made in 1086 and a succession of mills has operated on the site up to the end of commercial milling in 1974.

The mill derives its power from a slow moving stream which drains Pakenham Fen and becomes a tributary of the Black Bourn just below the mill. The great water-wheel is breastshot, water entering the rim buckets just above axle level. It is 16 ft in diameter and 8 ft 6 in. wide, typical of the wide wheels used in East Anglia to get maximum power from the comparatively shallow gradients of the rivers. The iron wheel was made by W. Peck of Bury St Edmunds whose name is cast on the wheel spokes and it was installed in 1902 to replace a wooden wheel which had served since 1814. A series of gears converts the turning moment of the main shaft to 120 rpm at the three pairs of millstones which are driven from below, a type of drive known as underdrift. To provide alternative power during times of drought a steam engine was installed during the 19th century and this machine was replaced about 1920 by a more flexible power source, a large Blackstone 21 hp oil engine. The Blackstone engine is still in the mill as is a rare 19th century Tattersall roller mill used to produce fine flour.

The four storey mill building, dated 1754, has a smart

appearance with a brick frontage added in 1810, a traditional weather-boarded lucam (hoist) projecting from the top storey, a pleasant fenestration and an unusual Georgian style ornate fanlight over the 'stone floor' door.

The Suffolk Preservation Society bought the mill in 1978 from the last miller, Mr B. F. Marriage (a famous name amongst East Anglian millers), and after renovating the building and machinery, opened it to public visits as a working water-mill.

28. Pakenham Windmill

HEW 839

TL 931 694

The parish of Pakenham is probably unique in having both a working windmill and a working water-mill (HEW 838) within its boundaries. The windmill stands in a farmyard about ½ mile south of Ixworth and less than ½ mile west of the water-mill.

The original date of the windmill is uncertain but it was built about 1830 for a Mr Fordham. The farm, with its mill, was bought in 1885 by Mr Bryant, whose descendants, John Bryant and his sons, still own it and have kept the mill in working order. Built in 'Wolpot White' bricks the tower is tarred externally for weatherproofing and has five storeys with a sack door and chute at first floor level. Small mullioned windows light each of the floors except the top level.

The rotatable cap, 17 ft in diameter, carries the four main 32 ft long sail stocks, two of which are fitted with patent sails. Pakenham was the only Suffolk tower mill known to have had a copper covered cap, but aluminium cladding was substituted in the early 1960s. The full height of the mill is 51 ft and the cap, surmounted by a 5 ft high wooden finial, carries a fantail and an external timber gallery with a neat crossed diagonal guardrail. The 12 in. thick upright shaft within the tower powers two pairs of 4 ft diameter French burr millstones, the drive being a combination of iron and hardwood cog teeth to eliminate sparking and thus reduce the danger of fire in the dust-laden air of a mill.

Suffolk was once rich in windmills and 643 corn mills have been recorded in the county. Now only half a dozen are preserved complete with machinery and Pakenham

Pakenham Windmill

is the last surviving working tower mill in private ownership in Suffolk.

Further reading

BROWN R. J. *Windmills of England*. R. Hale, London, 1976, 180–181.

29. Bury St Edmunds Station Bridge

HEW 923

TL 853 652

The railway from Ipswich to Bury St Edmunds, opened in 1846, was built by Joseph Locke with Peter Bruff as his resident engineer and Thomas Brassey the contractor, one of the most outstanding teams in railway construction of the day. An example of the high quality of their work is seen in the Station Bridge at Bury St Edmunds, for the design of which Locke retained Frederick Barnes, the Ipswich architect, who had already designed Bull Bridge at Brantham (HEW 901).

The slightly skew span of 54 ft 7 in. carries the railway over Northgate Street with an elegant semi-elliptical arch of seven courses. The brickwork is decorated by inset courses to give an ashlar block appearance and by dentilation below the parapet string course and the wing wall capping. The bridge was the central feature of a colourful ceremony on 7 December 1846 when a great crowd assembled to welcome the arrival of the Directors' train

OPENING OF THE BURY AND IPSWICH RAILWAY.—THE BURY STATION.

Opening of the Bury and Ipswich Railway, the Bury Station Bridge, from *The Illustrated London News*, 12 Dec. 1846

from Ipswich. The passenger service began on Christmas Eve and for nearly a year the bridge served as the terminus of the line until the adjacent station with its baroque towers and flamboyant cupolas had been completed.

Further reading

Illustrated London News. 20 June 1846, 377.

30. Moulton Packhorse Bridge

HEW 1716

TL 697 645

The little village of Moulton, 3 miles east of Newmarket has a packhorse bridge of considerable interest, thought to be the only remaining intact bridge of its kind in East Anglia. Dating from the 14th or early 15th century, it crosses the River Kennett which must have been wider at that time to have justified a bridge of this size. Alongside the bridge is a concrete paved ford on the minor road between Moulton and Gazeley which was once a trade route for packhorse trains between Bury St Edmunds and Cambridge.

The bridge is distinctly humped and the road is only 5 ft wide over the bridge, except at the ends where it opens out to a width of 10 ft. The four sharply pointed arches, lined with red brick, have nominal spans of 10 ft 6 in. and

Moulton Packhorse Bridge

are 7 ft 6 in. high. Short pointed cutwaters between the arches are edged with bricks and continue up the spandrels as rounded buttresses. The structure is faced with random knapped flints set in rough cement including the solid parapets which stand 2 ft 4 in. above the road.

The bridge is in remarkably good condition and its survival is evidently due to funds being available from an old charity which financed repairs for the fabric of the church and the bridge from medieval times. The bridge is now maintained by English Heritage and is a Listed Structure.

Further reading

WATKINS A. A. Moulton packhorse bridge. *Proc. Suffolk Inst. of Archaeology*. **XXI**, 1933, 110–119.

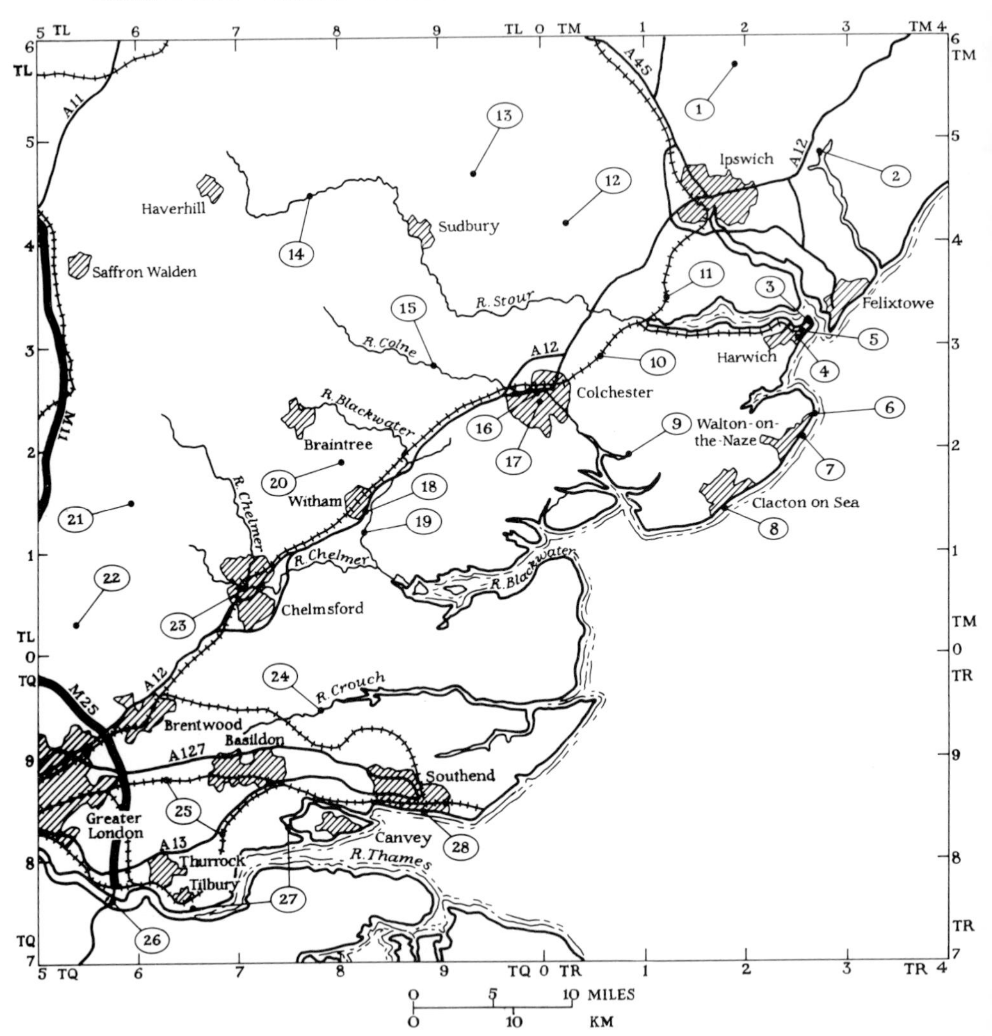

1 Helminham Hall Bridges
2 Woodbridge Tide Mill
3 Harwich lighthouses
4 Dovercourt Leading Lights
5 Harwich Treadwheel Crane
6 Walton-on-the-Naze Beacon
7 Walton-on-the-Naze Pier
8 Clacton-on-Sea Pier
9 Thorrington Tide Mill
10 London and Ipswich Railway
11 Bull Bridge, Brantham
12 Toppesfield Bridge
13 Brent Eleigh Bridge
14 Clare Bridge
15 Chappel Viaduct
16 North Bridge, Colchester
17 Colchester Water Tower
18 Saul's Bridge, Witham
19 Wickham Bishops Viaduct
20 Cressing Temple Barns
21 Aythrope Roding Windmill
22 Greensted Log Church
23 Stone Bridge, Chelmsford
24 Battlesbridge
25 London, Tilbury and Southend Railway
26 Queen Elizabeth II Bridge
27 Thames Estuary Tidal Defences
28 Southend Pier

5. South Suffolk and Essex

The southern half of Suffolk and the neigbouring County of Essex, which form the area covered by this section, reach from the chalk downs of Newmarket in the west to the fragile and convoluted coast of the Naze. London clay over much of the area has provided a ready base material for brickmaking in the absence of indigenous stone. There is a rich heritage of building crafts displayed in the thatched and pargetted houses of the towns and villages, epitomised in the splendour of Lavenham.

The craft of medieval carpenters is illustrated in several great barns, most notably the two barns of the Knights Templar at Cressing. An even earlier example of carpentry is the Saxon timber walled nave of Greensted church, the oldest timber church in the world.

The rolling countryside presents little demand for large engineering structures, although a great viaduct was needed to carry the railway across the Colne Valley at Chappel. At the other end of the size scale are three small cast iron bridges built by William Cubitt as a young engineer, his first being at Brent Eleigh in Suffolk.

Busy ports are an important feature of this seaboard, at Felixstowe, Ipswich, Harwich and Parkeston Quay. Along the coast, pleasure piers still survive at Clacton-on-Sea and Walton-on-the-Naze and Southend-on-Sea boasts the longest pleasure pier in the world.

In Victorian times the catalyst for large-scale development of southern Essex was the opening of the railways, which are now some of the most intensively used suburban railways in the country. In modern times, great engineering achievements in the region have been the River Orwell bridge at Ipswich, the Queen Elizabeth II suspension bridge at Thurrock and the tidal defence works along the Thames Estuary.

1. Helmingham Hall Bridges

HEW 341

TM 187 577

A pair of graceful and intricately decorated bridges cross the 60 ft wide moat which surrounds the great red brick mansion at Helmingham, the seat of Lord Tollemache, situated 8 miles north of Ipswich. The bridges are fine examples of the craftsmanship in iron casting achieved by Ransome, the Ipswich iron founder, and are among the earliest cast-iron spans still in use in East Anglia.

John Nash, the celebrated Regency architect, was retained to design major alterations to the Hall and his proposals with a strong Tudor-Gothic flavour were illustrated on drawings which he exhibited at the Royal Academy in 1800. One of his drawings shows a rather frail looking bridge of three equal cast-iron spans to replace the earlier brick arch bridge. The building alterations to the house were made between 1800 and 1803 and it is likely that the new bridges were erected during or shortly after that period. The bridge design was altered to a more substantial structure with a main span of 31 ft 6 in., semi-elliptical in profile and a side span with a pointed arch 9 ft wide to balance the span of the timber drawbridge at the inner side of the moat. The two drawbridges are still raised every night as they have been since 1510.

Main span, Helmingham Hall Bridge

The girders have vertically ribbed open spandrels with Gothic flourishes which are repeated in the decorated cast-iron balustrades. The three longitudinal girders are linked by cast-iron cross ribs with brick jack arches, supporting the brick paved deck. On the south-east bridge leading to the main entrance the road is 12 ft wide with a tar and gravel surface but the 8 ft wide deck of the north-east bridge still reveals the brick paving. The meticulous attention given to the overall balance in proportions of the structures, the decorative details on the ironwork and the ornate red brick octagonal pilasters produce an interesting complement to the facade of the Hall.

The bridges are private entrances to the mansion but they can be seen on the occasions when the Helmingham Hall gardens are opened to public visits.

2. Woodbridge Tide Mill

HEW 113

TM 276 487

The 200 years old tide mill on the River Deben at Woodbridge, a mill site since 1170, was the last operating tide mill in the country when it was taken out of service in 1957. The mill is a dominant feature at the north end of Woodbridge quay but the tidal pool which provided its head of water has been abandoned.

The millhouse has a brick base and a 23 ft high timber main frame, clad with horizontal boarding. The tiled mansard roof has a hoist on the landward side to enable sacks of grain to be lifted to the top floor. The 20 ft diameter water-wheel, carried on a 22 in. square oak shaft, powered four pairs of millstones.

After the wheel shaft broke in 1957 commercial use of the mill ceased and it was abandoned. During the subsequent 15 years it became progressively damaged and decayed. However, a local conservation group came to the rescue with support from local and national authorities and restoration work began in 1972. The base wall has been repaired and the main frame has been reconditioned, retaining much of the old timber. A replica of the original water-wheel has been fitted and the machinery and millstones reinstated and renovated so that demonstration runs of corn grinding can be made. The mill is open to visitors during the summer.

3. Harwich lighthouses

HEW 115

TM 261 324
TM 263 323

The old town of Harwich, sited on a narrow promontory where the Rivers Stour and Orwell reach the North Sea, has a distinctive tower standing alongside the main street, a 90 ft high lighthouse. This tower, together with the lower lighthouse on the edge of the shore, once provided a navigation bearing for vessels entering Harwich Harbour. The first navigation lights at Harwich, introduced in 1665, comprised a coal fire in a chamber over the Town Gate (now removed) and a lantern in a timber hut on the foreshore. John Constable illustrated the low lighthouse in a painting now in the National Gallery, London. These rudimentary lights survived for 150 years but, as a condition for renewing the licence in 1817, the owner, Major General Slater Rebow, was required to build two new lighthouses.

Work began in January 1817 and was completed in the following year, the new towers being built alongside and on the same bearing as the original lights. The two towers were designed by Daniel Asher Alexander and the work was supervised by John Rennie, Senior, then consulting engineer to Trinity House. The high tower built in grey gault brick has an unusual nine-sided polygonal plan and tapers from 20 ft 6 in. wide at the base to 13 ft wide at the top. The stone cap is crowned with a stone urn giving a total height of 90 ft. There are stone bands at three of the seven floor levels, the windows and doors have stone architraves and a wide stone canopy projects over the windows of the third storey. The lower lighthouse, built on the edge of the foreshore 220 yd from the high tower, is a shortened version of its lofty neighbour but with ten sides. It stands 45 ft high.

In 1837 the lighthouses were bought by Trinity House but the tenure of the new lights was to be short lived. As a result of a progressive extension of the shoal off Landguard Point the lights no longer provided a safe bearing for ships and when a number had run aground the lighthouses became known as 'misleading lights'.

On 2 November 1863 the lights at Harwich were extinguished and the navigation bearing was transferred to new leading lights at nearby Dovercourt (HEW 116). The High Lighthouse is now a private home and was restored

High Lighthouse, Harwich

in 1975 to mark European Architectural Year. The Low Lighthouse, an Ancient Monument, served for a time as a pilot station and is now leased to The Harwich Society as a maritime museum open to public visits.

Further reading

LONG N. *The lights of East Anglia*. T. Dalton Ltd, 1983, 137–156.

Dovercourt High Light Tower after restoration

4. Dovercourt Leading Lights

HEW 116

TM 253 308

The two iron light towers on the beach at Dovercourt were brought into use as navigating lights in 1863 when the lighthouses at Harwich (HEW 115), were taken out of service. The towers, probably designed by James Walker, consulting engineer to Trinity House, housed leading lights for vessels using the channel to Harwich Harbour.

They remained in use until August 1917 when the channel was marked with buoys, bringing to a close a period of 250 years during which leading lights had been displayed at Harwich.

Both towers stand on tubular cast-iron legs founded on iron screw piles, an invention of Alexander Mitchell, patented in 1830. The legs are inclined and the tower frames are stiffened below lantern level with solid iron plates.

The inshore tower, 56 ft high, has a six-sided lantern room 12 ft 6 in. in diameter and 9 ft high with bolted iron plate walls and a slate floor about 45 ft above beach level. Resistance to wave action is enhanced by diagonal and lateral bracing between the legs in the lower section of the tower.

The offshore light tower, 200 yds due east, is 41 ft 6 in. high with an octagonal lantern room 16 ft 6 in. in diameter and 8 ft high. This structure has four inverted A frame legs of cast-iron tubes and is wave swept at high tide. A rock and concrete causeway at beach level connects the two towers.

It is thought that Dovercourt Leading Lights are the last surviving pair of iron light towers on the British coast. They are classified as Ancient Monuments and a local conservation group, The Harwich Society is reconditioning the structures.

Further reading

LONG N. *The lights of East Anglia*. T. Dalton Ltd, 1983, 156–158.

5. Harwich Treadwheel Crane

HEW 1482

TM 262 325

Standing within a protective railing on Harwich Green is the only surviving example in the United Kingdom of a double treadwheel crane. It was constructed throughout in oak in the 17th century and served on the dockside for 260 years.

When hostilities against the Dutch brewed up in 1664 Harwich was well situated to serve as a base for the British fleet. The Duke of York visited the naval dockyard in 1666 and gave orders for improvements including the provision of a house crane similar to one then in use at Woolwich. In the following year, with the dockyard being expanded under the keen direction of Samuel

Pepys, the new crane was erected on the quay at a cost of £392.

The two treadwheels, 16 ft in diameter, 3 ft 10 in. wide and 4 ft apart on a common axle were operated by two men who walked inside the wheels. The lifting chain, winding on the 14 in. square wooden axle, with a 14:1 reduction ratio, passes along the jib and over a wheel. The jib projects 17 ft 10 in. and is 12 in. x 10 in. in section with a raking strut and a substantial kneeler. The jib is pivoted to swing through 180°. The house itself, which is really the crane frame, is 26 ft 3 in. x 14 ft 10 in. with stoutly braced 12 in. square main posts clad in weather-boarding. There was no brake on the crane and the inherent hazard this presented was slightly mitigated by a spar kept handy to jam against the edge of a treadwheel curb.

The crane continued to serve the dockyard up to the 1914–18 war but when the dockyard closed in 1928 the crane was given to the town. It was decided to move it to the Green and in 1930 F. Harold French supervised this operation, overcoming the problem of moving the wheels by transporting them on a temporary light railway. This Ancient Monument is preserved by the District Council with help from the Harwich Society.

Harwich Treadwheel Crane

6. Walton-on-the-Naze Beacon

HEW 749

TM 265 235

The beacon at Walton-on-the-Naze is an 81 ft high tower standing above the low cliffs on the blunt promontory of the Essex coast known as the Naze. It was built by Trinity House in 1720 as a daymark sited to line up with Walton Hall, some 350 yd further inland, as a guide to vessels passing through the Goldmer Gap in the shoals. This is the most massive and highest unlit beacon on the coast of the British Isles.

The octagonal brick tower, 18 ft 6 in. wide at the base, has three reducing sections, with clasping buttresses at the angles for the full height. Windows with arched heads are in the upper two sections, lighting the internal spiral staircase of 113 steps. The top of the tower was rebuilt in the 19th century possibly when a reinforcing framework was inserted at the top and the height reduced from the original 90 ft.

The tower is now about 80 yd back from the cliff edge but the history of erosion along this fragile coast points to the probability that the tower will eventually be threatened by the sea.

Walton-on-the Naze Beacon

Further reading

HAGUE D. B. and CHRISTIE. R. *Lighthouses, their architecture, history and archaeology*. Gomer Press, Llandyssul, 1975, 207, 211.

7. Walton-on-the-Naze Pier

TM 254 215

HEW 1659

The original pier at Walton, the fourth of our seaside piers, was built in 1830, the same year as the first Southend pier. It was 300 ft long, much shorter than the other pioneer piers, and it lasted for 65 years until it suffered severe damage in one of the storms characteristic of this exposed coast.

The importance to the resort of the holiday steamer traffic from London and Kent was such that the pier owner immediately set about replacing the structure. The new pier, 800 ft long on concrete piles, was built in 1895 by the contractor J. Cochrane and was later extended in several stages to its present length of 2600 ft. A railway, now removed, once ran along the pier to transport visitors.

At the shore end the pier is 150 ft wide with ten rows

of piles supporting the entrance and an amusement arcade. Vestiges of the earlier pier still remain under this section. The main axis of the pier is based upon simple bents of four piles with the outer piles raked to enhance lateral stability, concrete crossheads and a timber deck. The pier is privately owned by the New Walton Pier Co.

Further reading

BAINBRIDGE C. *Pavilions on the sea.* Hale, London, 1986, 208.

8. Clacton-on-Sea Pier

HEW 1658

TM 177 145

Clacton-on-Sea Pier was built in 1870–71 by Peter Bruff, the engineer and manager of the Eastern Union Railway, to enhance the attraction of Clacton as a seaside resort which had developed from a quiet fishing village in the short space of 20 years. The wooden pier was opened in 1872 as a passenger landing stage with a clear uninterrupted deck. The steamer traffic increased to such a level that the pier was extended to 1180 ft in 1893 to a design by Kinipple & Jaffrey and buildings were erected on the pier, including the ovoid pavilion and refreshment rooms at the seaward end. Later, the timber piles were replaced by concrete and the shore end of the pier was widened to

Clacton-on-Sea Pier

accommodate a vast amusement arcade, a lido and a fairground with a roller coaster and a Ferris wheel.

It is Britain's widest pleasure pier with a deck area of some 5 acres. The pier structure is based upon colonnades of concrete piles with a timber deck, except for the amusement hall area where it is concrete. At the seaward end of the pier an angled landing stage on timber piles has a concrete deck 3 ft below the general deck level. The pier is privately owned by Anglo-Austrian Automatics Ltd.

Further reading

BAINBRIDGE C. *Pavilions on the sea.* Hale, London, 1986, 191.

9. Thorrington Tide Mill

HEW 1579

TM 082 194

The Domesday survey recorded a number of tide mills in the convoluted estuaries and creeks to the north and south of the Naze but only two of this type of mill have survived in this area, at Thorrington and Woodbridge (HEW 113). The present mill at Thorrington was built in 1831 on the same site as the previous mills and the date is inscribed on a brick in the base of the building near the door. The mill stands at the head of Alresford Creek leading off the Colne estuary and at high tide, water was retained in the mill pound to the north of a raised farm track which passes the mill. On the ebb, water could be channelled from the pound to the external iron wheel, 16 ft in diameter and 6 ft wide, fitted with wooden paddles held in iron slots on the rim. The water-wheel is mounted on a very substantial cast-iron shaft 12 in. in diameter.

The four storey timber framed millhouse, 30 ft x 20 ft, is clad in white painted weather-boarding with a double pitched slated roof and has a timber lucam (hoist) on the east side. Inside the mill a 6 ft 6 in. diameter wooden pit wheel on the main shaft meshes with an iron wallower which drives the vertical wooden shaft. There are three pairs of 4 ft diameter millstones on the first floor.

The mill ceased operation in 1926 when the wheel failed, one family of millers, the Coopers, having worked it throughout the period 1841 to 1912. The last private owner was Tom Glover, farmer of the adjoining land, who did much to preserve the old mill until 1974 when it was bought by Essex County Council. The mill machinery

Thorrington Tide Mill

and millstones are intact and the County Council is restoring the mill to full working order.

Further reading

BENHAM H. *Some Essex water-mills*. Essex County Newspapers Ltd, 1976, 103–105.

10. London and Ipswich Railway

HEW 1662 (PART 1)

TQ 333 817 to TM 157 438

The pioneer railways in the North, Midlands and West of the country had already been built, or were under construction, when the Eastern Counties Railway (ECR) received parliamentary approval in 1836 to build the first railway into East Anglia, a line from London to Norwich and Yarmouth via Ipswich. The 115 mile route to Norwich was to take 13 years to complete, despite the simple geography of East Anglia, and would become the source of bitter rivalry between railway companies before it eventually became the main line.

Construction of the railway began in March 1837 with John Braithwaite as the engineer, working to a revised survey made in the previous year by Robert Stephenson and Joseph Locke, but progress was painfully slow. The long brick viaducts at the London end, the difficulty of crossing the Stratford marshes and a spell of unusually wet years delayed the work, and it was not until 20 June 1839 that the first section of line was opened between a temporary terminus at Mile End and Romford. Even then, ill fortune dogged the line as a train was derailed near Bow on the following day, killing the driver and fireman, the cause being excessive speed.

By mid 1840 the line had been extended to a new London terminus at Shoreditch and to Brentwood at the eastern end. Work continued eastwards including the deep cutting at Brentwood which required a gradient of 1 in 85, the most severe on the route. The service to Colchester finally opened on 29 March 1843. It had taken nearly 7 years to build the first 52 miles at a cost of £2.5 million compared with the original capital of £1.6 million for the whole route to Norwich and Yarmouth. The ECR could go no further. At Colchester, Samuel Morton Peto built the grandiose Victoria Station Hotel, designed by Lewis Cubitt in an Italianate style at a cost of £15 000, but competition from the established hotels in the town proved too much for the Victoria and in 1850 it was converted into an asylum, eventually being demolished in 1987.

Braithwaite had taken the eccentric decision to adopt a track gauge of 5 ft, although this had not been used elsewhere and the ECR soon experienced the problems of trans-shipping goods at junctions with other railways. Robert Stephenson advised changing to 4 ft 8½ in. gauge and this was accomplished during a hectic two months from September to October 1844. The ECR still had ambitions to reach Norwich and having taken over operations on the Northern & Eastern Railway to Bishops Stortford, extended the line from there to Cambridge and Norwich, opening the through service in 1845. Ipswich viewed with dismay the dual setback to the town's aspirations with the halting of the railway at Colchester and the ominous arrival of an alternative route at Norwich. Fearing they would be isolated, a group of Ipswich businessmen launched the Eastern Union Railway to build a line linking up with the ECR at Colchester.

The 17 mile extension to Ipswich was designed by Joseph Locke. Peter Bruff was the Resident Engineer. Timber piled viaducts were built over two arms of the Stour estuary at Cattawade (TM 102 326) and a tall five-arch brick bridge was erected at Brantham (HEW 901). The new line was opened in June 1846 using a temporary station at Ipswich until the tunnel had been built, when the present station site was developed.

Further reading

GORDON D. I. *Regional history of the railways of Great Britain; the eastern counties*. David and Charles, Newton Abbot, 1977, Chapters 2 & 4.

Bull Bridge, Brantham

11. Bull Bridge, Brantham

HEW 901

TM 121 345

North of the Stour estuary the Eastern Union Railway between Colchester and Ipswich required 1 mile long cutting near the village of Brantham, reaching a depth of 60 ft. At its deepest point the cutting is crossed by the A137 Colchester–Brantham–Ipswich road, which is carried over the railway by an imposing five-span brick bridge known as Bull Bridge, a name derived from a nearby inn.

Joseph Locke, the engineer, selected an Ipswich architect, Frederick Barnes, to design Bull Bridge. The five semi-elliptical arches, of about 29 ft span are carried on slender piers which are unusual in being buttressed for the full height to parapet level. Although the bridge was built mainly in red brick, Barnes chose to decorate the structure with 'Ipswich white' brickwork in the pier buttresses, arch faces and parapet string courses. The bridge still retains its clean-cut functional appearance, slightly marred by some random patch repairs in non-matching brickwork.

Further reading

Illustrated London News, 20 June 1846.

12. Toppesfield Bridge

HEW 837

TM 026 421

On the west side of Hadleigh a minor road crosses the River Brett over a medieval three-span bridge built in brick and stone, described by Jervoise as the 'finest ancient road bridge in Suffolk'. The bridge is of special interest historically as an illustration of several bridge-building periods and techniques.

The original bridge probably dates from the 14th century and was built from heavy stone blocks forming pointed voussoir arches with spans of 10 ft. The tapered stone blocks were assembled as six arch ribs in each span and the intervening spaces were then covered with stone slabs. This system of stone ribs is still to be seen in the downstream side of the arches. Random stone blocks in the downstream face indicated that the spandrels were originally built in that material and the remaining brickwork was probably part of restoration work carried out in 1640.

As Hadleigh grew and the country traffic increased a wider bridge was needed, and in 1750 the stone arches were extended in brick work on the upstream side. In the 1960s the County Council discovered that the soffit slabs had deteriorated, making the bridge inadequate for modern road traffic. Strengthening work began in 1974, involving the removal of all the fill material down to the arches and the provision of a reinforced concrete slab across the structure on raised piers and abutments, thus relieving the load on the arches. This added yet another form of construction to the bridge. Despite the changes in its construction over a period of some 600 years, the classic lines of the bridge have been retained. The brick parapets have a moulded ridge capping with octagonal capped pilasters at the four corners. The bridge is classified as an Ancient Monument.

Further reading

JERVOISE E. *The ancient bridges of mid and eastern England.* The Architectural Press, 1932, 128.

13. Brent Eleigh Bridge

Brent Eleigh Bridge, South face

HEW 1642

TL 934 483

In 1812 William Cubitt, a young millwright from Horning in Norfolk, joined Ransome & Son the Ipswich agricultural implement makers and began a civil engineering career which was to take him to the top of his profession, Presidency of the Institution of Civil Engineers and a knighthood. The opportunity to build his first road bridge came in the following year with the replacement in cast-iron of a small span carrying the Hadleigh to Bury St Edmunds road over the River Brett about ½ mile west of Brent Eleigh village.

Cubitt chose a semi-elliptical profile for the 13 ft wide arch of seven cast-iron ribs with a rise of 3 ft. A lateral rib links the units at the crown of the arch and through bolts are fixed at the mid point of the spandrels. Cast-iron bearing plates on the brick abutments (which are probably remnants of the previous bridge) support the ribs and the soffit of the arch is lined with timber planks between the ribs to contain the concrete fill. On the cast-iron face units, raised figures show the date of 1813.

The bridge cost £648 and was quickly followed by two more bridges of similar style and decorative pattern which Cubitt built at Clare, Suffolk (HEW 1689) and Witham, Essex (HEW 1362). In 1953 the road was diverted past Cubitt's bridge which was left *in situ*. The bridge is classified as an Ancient Monument.

14. Clare Bridge

HEW 1689

TL 767 449

The second of William Cubitt's cast-iron bridges carries a country road over the River Stour at Clare, an historic town with a ruined castle and a 13th century monastery. Early in 1813 William Brown, the County Surveyor, found that the earlier bridge at this site was beyond repair and turned to Ransomes at Ipswich for the new type of cast-iron superstructure which had already been designed by Cubitt for the single span at Brent Eleigh (HEW 1642).

The bridge at Clare is a more elaborate structure of three spans but the pattern of the Brent Eleigh bridge is repeated. The semi-elliptical arched spans of 11 ft, 13 ft 6 in. and 11 ft each have seven cast-iron ribs. These are tied at the crown and through bolted at mid spandrel level. Both faces of the bridge bear the cast date of 1813 in raised figures. An unusual structural feature is an extra cast-iron rib fixed to the west face and supported on cantilevered brackets giving a tapered widening of the deck from the mid point of the bridge. The reason for this partial widening is not documented and it appears to have been a late amendment to the design. The parapet railings end in short brick walls which have cast-iron caps marked Ward & Silver, Melford. The bridge is in remarkably good condition despite exposure to heavy vehicles and occasional high flood levels.

15. Chappel Viaduct

HEW 593

TL 895 282 to TL 897 286

Chappel Viaduct, a Listed Grade I Structure, is the longest bridge in East Anglia and makes an impressive sight with 32 high arches striding across the Colne Valley above the hamlet of Chappel, formerly known as Chapple. Peter Bruff, who was prominent in constructing railways in East Anglia, was the engineer for the Stour Valley Railway and built the viaduct in the period 1847 to 1849 at a cost of £21 000. The structure is 1136 ft long with semicircular arches of 30 ft span and reaches a

Chappel viaduct

maximum height of 74 ft above the River Colne. The seven million bricks used in the construction were made by the contractor, George Wythes, at a brickworks he set up less than 1 mile from the site. The somewhat austere appearance of the viaduct is relieved by slotted piers and corbelled imposts below the arch springing level. An interesting stone tablet set into the face of Pier 21 records the start of work on the viaduct on 14 September 1847. Bruff described the design and construction of the viaduct in a paper read to the Institution of Civil Engineers on 26 March 1850 and stated that cost had influenced his preference for building the arches in brick rather than laminated timber as originally intended. In the discussion which followed the paper Isambard K. Brunel, an advocate of timber spans which he had used extensively in the West Country, firmly refuted the view that there was cost advantage of brick compared with timber spans. Bruff was greatly pleased with his viaduct and commissioned Frederick Brett Russell RA to paint a landscape of the structure crossing the valley. The painting is now in the Ipswich Museum.

The viaduct was wide enough to carry two tracks, but only one was laid as the expected growth in rail traffic did not materialise and the widespread rail system in the Essex–Suffolk border area was eventually dismantled, leaving only the short Sudbury branch line which crosses the viaduct.

Further reading

BRUFF P. Description of the Chapple Viaduct, upon the Colchester and Stour Valley extension of the Eastern Counties Railway. *Proc. Instn Civ. Engrs. 1850*, **9**, 815.

16. North Bridge, Colchester

HEW 1595

TL 994 257

A crossing of the River Colne at the site of North Bridge has probably existed since Colchester was the principal Roman city in East Anglia. The date of the first bridge here is unknown but a bridge is depicted on Keymer's map of 1648 during the bitter siege of Colchester in the Civil War. Jervoise refers to a timber bridge which existed in 1748 and this was rebuilt in three brick arches by Sir William Staines, Lord Mayor of London, early in the 19th century. Within the short space of 40 years this structure became unstable and the Borough Corporation, alarmed by the collapse of another bridge over the Colne on the east side of the town in 1839, decided to rebuild North Bridge.

The new bridge was constructed in 1843 with three spans of cast- iron girders, the centre arch of 30 ft 6 in. and two side arches of 17 ft. The roadway and footpaths had a total width of 31 ft which was generous for the time but this was to prove inadequate for such a busy thoroughfare and the bridge was widened to 48 ft on the east side in 1903. H. Goodyear, the Borough Engineer and Surveyor, carried out the work and retained the original cast-iron face ribs and the cast balustrade. Plaques on the parapets record the original construction of the bridge in 1843 and the widening in 1903.

17. Colchester Water Tower

HEW 1688

TL 993 252

Colchester water tower on Balkerne Hill has been one of the most distinctive features of the town skyline since 1883. At that time it was an object of public controversy but in later years it has become a matter of civic pride with the local name of 'Jumbo'. The massive tower, designed by Charles Clegg the Borough Surveyor and Engineer, was erected by a local builder Henry Everett, and 1 200 000 red bricks were used in the construction. On each side of the tower, high arches are faced with Corsehill stone and the top has decorative crenellation. A central shaft carries the rising and supply mains. The tank, made of cast-iron plates cross tied with heavy iron bars, is 55 ft 7 in. square by 12 ft deep with a capacity of 221 000 gallons. The roof is surmounted by a wooden lantern and at a height of 131 ft the tower dwarfs the surrounding buildings. Originally the roof of the tank was tiled but the cladding was changed in 1948 to copper sheets overlaying felted boarding.

The tower was built on this site to draw from an artesian well drilled by Peter Bruff, the railway engineer, but the Borough water supply now comes from distant reservoirs and 'Jumbo' is used only as a balancing tank for the daily peak loading.

Colchester Water Tower

18. Saul's Bridge, Witham

HEW 1362

TL 824 139

Above: Saul's Bridge, Witham

The oldest cast-iron bridge still in use in the county of Essex, Saul's Bridge at Witham, was built in 1814 at a cost of £700. It was the third bridge designed by William Cubitt. The cast-iron ribs are similar in pattern to those of his two earlier bridges in Suffolk, at Brent Eleigh (HEW 1642) and at Clare (HEW 1689). The single span of 18 ft carries the B1018 Witham to Langford road over the River Brain, at this point just a stream. The two outer ribs cast marked 'RANSOME & SON IPSWICH 1814' and seven inner ribs hold soffit plates also made of cast-iron with a concrete infill below the road surface. Despite the shallow construction depth the bridge continues to carry a busy traffic flow. The bridge had no footpaths and with the relentless growth in road traffic it became necessary in 1955 to construct a concrete footway on the west side, but the original cast-iron parapet railings have been retained on the other side of the road.

19. Wickham Bishops Viaduct

HEW 1723

TL 824 118

At Wickham Bishops a remnant of the dismantled railway between Witham and Maldon is thought to be the only surviving railway timber viaduct of the 19th century

Wickham Bishops Viaduct

in England, although there are still a few examples in Wales and Scotland.

The railway was built by Joseph Locke with Thomas Jackson as the contractor and the line was opened for passenger trains on 2 October 1848. To cross the two arms of the River Blackwater at Wickham Bishops, Locke built a timber viaduct 130 yds long, carrying two tracks. Later, during the period that the line was part of the Great Eastern Railway system, a short embankment was substituted for the section of structure between the two streams, effectively splitting the viaduct into two parts. This change was made to improve stability because of the excessive vibrations generated by trains crossing the structure. Probably at the same time the viaduct was reduced to carry a single track. The two sections have a total of 21 spans with single trestle bents of 12 in. x 12 in. timber and longitudinal bearers of similar size.

The railway closed on 15 April 1966 and the track was lifted in 1969. Since that time the viaduct has become much decayed and partly overgrown with trees. Essex County Council took over the disused railway route in 1993. As the viaduct is a Scheduled Ancient Monument it is hoped that the structure will be preserved.

20. Cressing Temple Barns

HEW 1821

TL 800 187

Agricultural barns are not usually expected to rank as prime examples of engineering but the two great barns at Cressing Temple near Braintree have outstanding merit as original specimens of early timber structures and illustrate in a spectacular way the skills of medieval carpenters. Said to be the finest pair of medieval barns and the largest early timber framed buildings in Europe, these barns are deservedly Listed Grade I.

Built about 1260, the Wheat Barn is 130 ft long, 44 ft wide and 40 ft high with a tiled roof, hipped with gablets at both ends.

The Barley Barn, which preceded the Wheat Barn by some 40 years, is marginally shorter in length and has a similar style of roof covering. Both barns have a large projecting porch (midstrey) at mid length, of a size capable of admitting a loaded cart. The interiors have something of an ecclesiastical air, open to the lofty roof ridge, with roughly 12 in. by 12 in. main posts flanking the central nave. The oak roof structures are of special interest, with their intricate but economical arrangements of braced struts and ties. The Wheat Barn roof retains the

Top: Cressing Temple Barn

original system of passing braces and intermediate trusses whereas the Barley Barn was modified to a crown frame in the 16th century. There is plenty of evidence of the notched lap joints favoured by 13th century carpenters and it is estimated that they had to make about 1200 joints in constructing the framework for each barn.

The site has an intriguing history. Cressing Manor was granted to the Order of the Knights Templar in 1137 by Queen Matilda, wife of King Stephen. The Order was first established in Jerusalem, based upon the el Aksa Mosque, which the Knights identified as Solomon's Temple, and from there they carried out their self imposed task of protecting pilgrims to the Holy Land. During the tutelage of the Knights Templar at Cressing the barns were built to take in the harvest so that the grain could be threshed under cover during the winter.

After the influence of the Templars declined the manor remained in private ownership until 1987 when it was acquired by Essex County Council. Since then, the Council has undertaken extensive repairs to the barns and the site is open to public visits at weekends.

21. Aythorpe Roding Windmill

HEW 1596

TL 590 152

Essex County Council, which was the first local authority to acquire and maintain windmills as historic structures,

Below: Aythorpe Roding Mill and Farm

has an unusually large post mill at Aythorpe Roding in the heart of the rural district of the Rodings. This has been a windmill site for many centuries and, although there is no record of the exact date of this mill, it was indicated on a map of 1766. Thomas Belsham became the miller in 1892 and his two sons continued to work the mill until it ceased operating in 1935.

The rotating wooden body of the mill (the buck), 12 ft x 20 ft and 45 ft to the top of the aluminium covered cap, is supported on a massive timber main post 30 in. thick. Two pairs of 4 ft diameter French burr millstones are driven by four double shuttered patent sails and a later addition on the 'stone' floor is a flour dressing machine called a bolter. The 12 ft diameter fantail mounted on the steps to the buck, with a wheeled tailpost running on an annular track, keeps the sails facing into the wind. The brick roundhouse, 24 ft in diameter, was probably a later addition. Within the roundhouse is a third set of millstones for steam power installed in the late 19th century bearing the name 'F. Christy — Millwright, Chelmsford.'

After being disused for 40 years the mill had by 1975 become decayed and unstable. The County Council's own millwright Vincent Pargeter took on the task of restoring it over a period of seven years at a cost of £78 000. It was no small triumph after the painstaking work of repairing the mill, using wherever possible the old timbers, that flour was produced again on 3 March 1982 after a lapse of almost half a century. The mill is open to public visits at certain times.

Further reading

BROWN R. J. *Windmills of England.* Robert Hale, London, 1976, 68–69.

22. Greensted Log Church

HEW 1475

TL 539 030

The hamlet of Greensted, 1 mile west of Chipping Ongar, has a unique building, the only surviving example of a Saxon timber walled church which is reputed to be the oldest timber church in the world. The walls of the nave consist of vertical half logs of oak which have been dated at about the middle of the 9th century.

The nave illustrates an interesting development in building techniques between the 7th and 9th centuries.

Archaeological explorations below the church floor in 1960 revealed evidence that the walls of an earlier Celtic building consisted of upright logs placed directly into a trench. Two hundred years later the Saxons used a more advanced method of building their walls by fashioning tenons at the base of the logs which fitted into a slotted timber sill. The log walls remained in their original form for 1000 years until 1847–48 when the bottom of the logs had become so decayed that they were shortened and supported on a shallow brick wall. The inner faces of the logs still show the adze marks of the Saxon carpenters.

This heritage gem remains in regular use for religious services and attracts many visitors, particularly overseas tourists. The Post Office featured the church on the 3p postage stamp issued in June 1972.

Further reading

PEVSNER N. *The Buildings of England. Essex.* Penguin, London, 1965, 215.

23. Stone Bridge, Chelmsford

HEW 1368

TL 710 065

The original ford across the River Can was first bridged in 1100 when Chelmsford was a tiny settlement of five cottages. The second bridge on this site was built in the mid 14th century at the behest of Ralph, Bishop of London, the work being carried out by Henry de Yeucle, the master builder of Westminster Abbey nave. This medie-

Stone Bridge, Chelmsford, west face

val bridge had three pointed arches with two piers in the river and it lasted until 1788 when the present bridge, designed by the County Surveyor John Johnson, was opened.

Stone Bridge, standing at the end of the High Street, has a single segmental arch of 35 ft 9 in. span built in Portland stone with attractive balusters of terra cotta. The keystone on the west face is inscribed 1787 and the spandrels are decorated with stone plaques depicting river gods. The bridge is a Scheduled Ancient Monument.

Evidently the two piers of the medieval bridge were left standing in the river at the time the Stone Bridge was constructed and were later suspected of contributing to the serious floods which affected the town. Thomas Telford was consulted in 1824 and recommended clearing and deepening the river. This led to the removal of the offending piers and some alleviation of the flooding problem. The river-banks are now walled through the town.

Further reading

JERVOISE E. *The ancient bridges of mid and eastern England*. 1932, 134.

24. Battlesbridge

HEW 1476

TQ 781 947

The first recorded mention of a bridge over the River Crouch at this site was in 1327 when its name was given as Batailesbrigge, most probably by association with the local family of Bataille. Records indicated that the timber trestle bridge was in constant need of repair and one statement in the Sessions Records of 12 July 1571 during the reign of Elizabeth I said 'a sertayne brydge called Battells brydge....is ruynouse and in great decay'. However, the bridge remained until 1855 when it was replaced by a cast-iron span of 97 ft 6 in. This only lasted until 1871 when it was destroyed by a steam traction engine, stated in a local newspaper at the time to be 'a monster of 12 tons'.

A new bridge with one main span of cast-iron girders and two brick side arches was designed by the County Surveyor, Henry Stock and in 1872 the bridge was built at a cost of £3500 by William Webster of St Martins Place, Trafalgar Square. The bridge crosses the river just below the tidal limit and, with a tidal range of 6 ft at ordinary tides, substantial piers 10 ft wide were needed. The 47 ft

Battlesbridge

6 in. main span consists of a gently curved segmental arch of six cast-iron ribs spaced by brick jack arches. The ribs were cast in three sections and assembled with flanged bolted joints. The east parapet still has the original cast-iron openwork fence with a geometric pattern and the pilasters are stone capped in a Flemish style.

It says much for the strength of the cast-iron span that it survived over a century of increasingly heavy traffic along the A130 Chelmsford to Benfleet road until a by-pass was constructed. Unfortunately, this diversion did not occur in time to save the bridge from the indignity of an incongruous concrete footbridge being erected in 1958 on the west side.

25. London, Tilbury and Southend Railway

HEW 1687

TQ 335 810 to TQ 940 850

The London, Tilbury & Southend (LTS) Railway, running east from London to Southend along the north side of the Thames, was built in 1854–56 to provide a service between London and Tilbury, where the ferry across the Thames had just been started, and also to Thames Haven which had possibilities as a fish and cattle port. The extension out to Southend was a secondary objective. In the event, this generated a vast expansion of population in South-East Essex, particularly in the Southend area which became one of the largest coastal resorts in the country, and the railway developed into a major commuter route into the City.

The LTS was authorised in June 1852 and was designed by G. P. Bidder; the contract was awarded to the firm of Peto, Brassey & Betts. In April 1854 the double line from Forest Gate Junction via Barking reached Tilbury. It was not until 1886 that the original aim of the line could be fulfilled with the £3 million development of Tilbury Docks by the Port of London Authority. In 1854 construction continued eastwards to Pitsea and ran just above high tide level at Leigh, reaching Southend in 1856. The Southend railway beyond Tilbury was a single line without the benefit of telegraph and until it was doubled only a few trains could be run each day. LTS trains travelled to the Eastern Counties terminus at Bishopsgate via Forest Gate. This uneasy arrangement continued until 1858 when the LTS obtained powers to run into Fenchurch Street station over a newly built connection from Gas Factory Junction to Barking.

East of Barking the Tilbury loop turns south and the more direct route to Southend, opened in 1888, passes through Upminster. This new part of the route was built by Arthur Stride, the Engineer and General Manager of the LTS, to compete with the Great Eastern Railway alternative service to Southend from Shenfield. Stride had also extended the LTS to Shoeburyness in 1884.

The Tilbury Loop was to play an important part in the enormous industrial development of South-East Essex. At Dagenham the Ford motor plant is served by the line. Tilbury still has a passenger station, but the great fans of sidings once totalling 30 miles of track have mostly disappeared. On the eastern leg of the Loop the branch line to Thames Haven now serves the huge oil terminals. The Midland Railway took over the LTS in 1912 with a promise to electrify the line to Southend, an aim which was not achieved until 50 years later.

Further reading

WELCH H. D. *The London, Tilbury & Southend Railway*. Oakwood Press, London, 1963.

26. Queen Elizabeth II Bridge

HEW 1917

TQ 573 763

The impressive suspension bridge across the River Thames at Dartford is a striking example of modern civil engineering. It is presented as an historical engineering work despite its recent date because it represents an important advance in major bridge technology. At the time it was opened in October 1991, it was the longest cable-stayed span in Europe.

The steel and concrete masts, rising 137 m above the massive piers, support the deck with 56 pairs of cables. The 450 m river span is part of the 812 m long cable-stayed deck. Beyond the river structure are 1 km long viaduct sections to the north and south, linking up with the existing road system. The bridge was designed by Trafalgar House Technology and constructed by the Cleveland Bridge Co.; Cementation Construction built the mast foundations.

The bridge carries four lanes of southbound traffic on the A 282 link in the London orbital motorway (M25); the northbound traffic uses the adjacent twin tunnels.

Further reading

BYRD T. Breaking the bottleneck. *New Civil Engineer*, 6 June 1991, 24–31

WATSON R. Open for business. *New Civil Engineer*. 31 Oct. 1991, 24–28.

Above: Queen Elizabeth II Bridge

27. Thames Estuary Tidal Defences, South-East Essex

HEW 1706

TQ 54 78 to TQ 93 84

The great tidal surge up the Thames Estuary on 31 January 1953 breached the flood banks and inundated large areas of South-East Essex. During the night the sea burst through the defences at Canvey Island drowning 58 people. Twenty-five thousand people were rendered homeless and Central London escaped flooding that night by the narrowest of margins. The studies which followed this disaster projected a remorseless rise in tide levels and Parliament authorised the construction of the Thames Barrier at Woolwich to protect London, together with the strengthening of the estuary banks downstream.

The strengthening of the Essex banks from Purfleet to Leigh-on-Sea included raising the crest with reinforced concrete walls or deep steel sheet piling, but more elaborate and innovative measures had to be adopted at special locations. The entrance to Tilbury Docks was installed with a retractable flood gate 110 ft long and 59 ft high. The creeks which surround Canvey Island have to remain open during normal conditions and flood protection has been provided by three flood barriers sited at Fobbing Horse, (TQ 740 844), Easthaven (TQ 748 843) and Benfleet (TQ 782 855). The largest of these, across Vange Creek at Fobbing Horse, has a clear span of 110 ft with a 1600 t beam supporting the winch house and a 1200 t hinge beam holding the four steel gates.

This Anglian Water project was carried out between 1974 and 1982 at a cost of £104 million. The scheme was designed by Binnie & Partners and the contract was placed with John Howard.

Further reading

New Civil Engineer. 15 November 1979, 27.

Thames Barrier and Flood Protection Act 1972.

28. Southend-on-Sea Pier

HEW 79

TQ 885 850 (landward end)

The first pier at Southend, opened in 1830, was a timber structure 1500 ft long, built to attract pleasure steamer traffic to the small but growing resort. The pier was extended by James Simpson in 1845 but the wooden piles

Southend-on-Sea Pier

were vulnerable to attack by marine borers and it was replaced by the present pier in 1889–91. The new pier, 6400 ft long, is supported on trestle bents of cross-braced cast-iron columns with cast-iron screw piles (probably Mitchell's type). It was designed by James Brunlees and built by Arrol Bros of Glasgow at a cost of £80 000. The pier head, designed by Brunlees and J. W. Barry was completed in 1898 and further extended in 1927 making the total length an impressive 7080 ft (1.34 miles). Southend has retained to this day its reputation of having the longest pleasure pier in the world.

The pier has suffered most of the indignities visited upon seaside piers: devastating fires, structural deterioration, mounting maintenance costs, declining revenues and damage on many occasions from colliding vessels. In the 1970s it was doubtful whether or not the pier could survive but with energetic local support and help from the Historic Building Fund, Southend Borough Council launched a programme of refurbishment giving the pier a fresh lease of life. Included in the new facilities are diesel powered trains on a 3 ft gauge single track railway, with a passing loop, thus continuing the history of tram and train transport along the pier dating from the 1870s.

Further reading

BAINBRIDGE C. *Pavilions on the Sea*. Robert Hale, London, 1986, 204–205.

ADAMSON S. H. *Seaside Piers*. B. T. Batsford, London, 1977, 107.

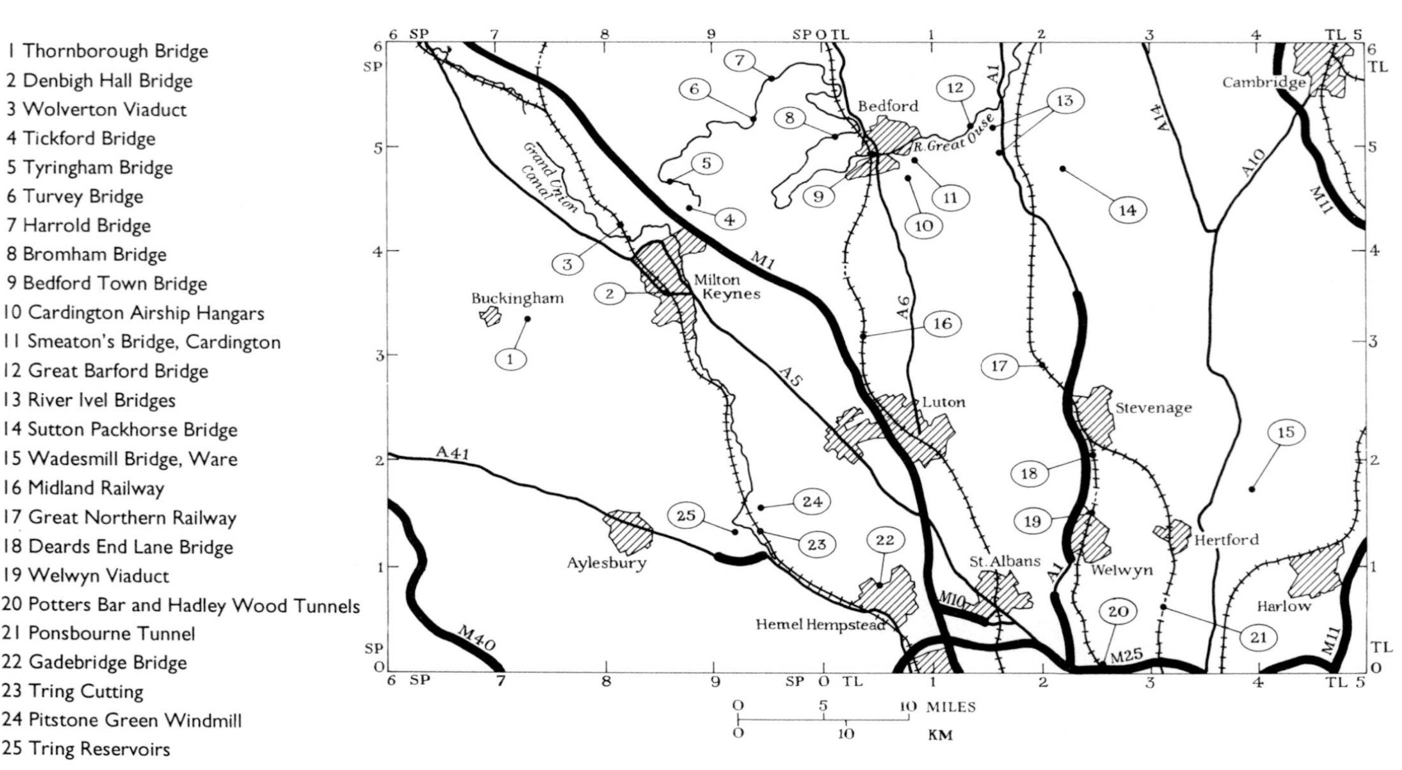
1 Thornborough Bridge
2 Denbigh Hall Bridge
3 Wolverton Viaduct
4 Tickford Bridge
5 Tyringham Bridge
6 Turvey Bridge
7 Harrold Bridge
8 Bromham Bridge
9 Bedford Town Bridge
10 Cardington Airship Hangars
11 Smeaton's Bridge, Cardington
12 Great Barford Bridge
13 River Ivel Bridges
14 Sutton Packhorse Bridge
15 Wadesmill Bridge, Ware
16 Midland Railway
17 Great Northern Railway
18 Deards End Lane Bridge
19 Welwyn Viaduct
20 Potters Bar and Hadley Wood Tunnels
21 Ponsbourne Tunnel
22 Gadebridge Bridge
23 Tring Cutting
24 Pitstone Green Windmill
25 Tring Reservoirs
Cambridge
Bedford
R. Great Ouse
Grand Union Canal
Milton Keynes
Buckingham
Luton
Stevenage
Hertford
Harlow
Welwyn
St. Albans
Hemel Hempstead
Aylesbury
M1
M10
M11
M25
M40
A1
A5
A6
A10
A14
A41
0 5 10 MILES
0 10 KM

6. Hertfordshire, Bedfordshire and North Buckinghamshire

The three counties covered in this section straddle the main traffic routes between London, the Midlands and the North. Within the region, a major river, the Great Ouse, flows across these routes in a north-easterly direction. It is normally a placid stream but its propensity to flood over a wide, soft flood plain has, over the centuries, often affected communications. A number of items in this section illustrate the continuing task of building and maintaining bridges to overcome the restraint to travel presented by this capricious river.

The gentle terrain of the Chilterns has not, with some notable exceptions, required the construction of large viaducts and tunnels. However, the scarp did present a significant barrier to the early engineers who built roads, canals and railways across this line of hills.

The construction of the Grand Junction canal provided the first major route in the area for the bulk transport of freight between the industrial district of Birmingham and the capital. The route followed the narrow valleys of the Rivers Bulbourne and Gade, and crossed the scarp by a combination of deep cutting and a flight of locks. The London and Birmingham railway took the same route, with a deep cutting little more than ¼ mile away from that of the canal. Other railways crossed the region, the Great Northern, the Midland, the Great Western line to Birmingham and lastly, the Great Central.

Many distinguished engineers have been associated with the development of these national transport routes. However, there are some fine structures, particularly bridges over the River Great Ouse, which were built by local craftsmen, some of whom remain unknown, whose works stand as durable marks of their skill.

1. Thornborough Bridge

HEW 175

SP 729 332

A long narrow causeway crosses the valley of the Padbury Brook close to the meeting point of several ancient tracks and earthworks. Six spans, of between 10 and 11 ft, are separated by massive piers with cutwaters.

Said to date from the 14th century, the bridge formerly carried the A 421, Bletchley to Buckingham Road. The road over the bridge is only 11 ft wide but the triangular cutwaters are carried up to road level to form refuges, and a wide rectangular projection on the north face was probably intended for the same purpose. The Quarter Sessions lists a number of occasions on which repairs had been undertaken and a plaque mounted on the north face records such an occasion in 1661.

The bridge was bypassed to the south in 1974, the old road forming a useful parking area from which the bridge may be crossed on foot. The new bridge offers a useful vantage point for a good view of the old bridge.

Further reading

JERVOISE E. *The ancient bridges of England and Wales*. The Architectural Press, London, 1932, 83–84.

Above: Thornborough Bridge

2. Denbigh Hall Bridge, Milton Keynes

HEW 415

SP 863 353

In April 1838, the London and Birmingham Railway was opened as far north as Denbigh Hall, where the line crossed Watling Street (A5) between Fenny Stratford and Stony Stratford. Owing to construction difficulties with Kilsby tunnel (HEW 55, chapter 7), that section of the line had not yet been completed. Denbigh Hall formed a temporary terminus at which rail passengers from Euston were transferred to stage coaches for the journey to Rugby before completing their journey to Birmingham by rail. A plaque on the south-east portal of the bridge commemorates the opening of the completed line in September 1838.

A comparison of a contemporary print with the largely unaltered abutments indicates that the heavy skew was accommodated by lengthening the walls and placing the deck girders square. Whereas the original girders had curved soffits, straight girders were used when the line was quadrupled in 1892. The width of the bridge is 165 ft and the square span 26 ft 9 in. All the cast iron girders under the tracks were replaced with concrete beams in 1946, except for five single units and a double fascia beam at the northern end. There are, therefore, examples of the complete history of the deck construction within the existing bridge.

Denbigh Hall Bridge, Milton Keynes

3. Wolverton Viaduct

HEW 152

SP 815 422

To carry the line of his new London and Birmingham Railway across the River Great Ouse in 1838, Robert Stephenson designed a brick viaduct which carries the line 50 ft above the river. There are six main spans with semi-elliptical arches of 60 ft span; the viaduct was originally 22 ft wide. The piers are 10 ft 10 in. wide increasing to 11 ft 3 in. at the bottom and the viaduct terminates in massive solid piers 28 ft long. Behind these piers is a hollow abutment pierced by semi-circular brick arches of 15 ft span, springing from piers 6 ft wide. These piers are in turn pierced by openings about 6 ft wide. This form of abutment construction appears to be common to most of Stephenson's viaducts on the London & Birmingham Railway and is an early example of the standardisation of design.

In 1882 the viaduct was widened by adding an extra 31 ft on the west side in order to allow for the quadrupling of the tracks, giving a total width of 53 ft. The new viaduct is similar to the original 1838 structure and the joint between the two structures can be clearly seen, the two structures not being bonded together.

The main features of the viaduct are well illustrated in a contemporary engraving of the 1838 structure.

Further reading

STEEL W. L. The History of the London & North Western Railway. *The Railway and Travel Monthly*, London, 1914, 73.

Above: Wolverton Viaduct

4. Tickford Bridge, Newport Pagnell

HEW 52

SP 877 438

Built in 1810, this cast-iron bridge is still capable of carrying vehicles up to the present day maximum loading, albeit after some strengthening.

The bridge is 25 ft wide and spans 58 ft across the River Ouzel (or Lovat) between limestone masonry abutments which are carried down to bedrock. The superstructure is formed by six equally spaced cast-iron ribs connected by cast-iron transverse diaphragms. Each rib is composed of eleven segments with mortice and tenon joints keyed together to form an arch. Spandrel panels consist of iron rings diminishing in size towards the crown. Iron deck plates are placed over the upper member of the spandrel panel and located by continuous longitudinal lugs. The bridge was jointly designed by Thomas Wilson and Henry Provis, following Wilson's patented designs. The iron work was cast at the Walker Foundry, Rotherham, Yorkshire and construction was under the control of Provis.

Following the fracture of one of the deck plates in 1900 the three inner bays were strengthened by the addition of wrought iron arch plates, bolted through the iron deck plates close to the ribs. In 1976 a reinforced concrete slab was laid over the deck on a ¾ in. thick cushion of plastic foam.

Tickford Bridge, Newport Pagnell

There is no footpath over Tickford bridge; the footpath uses an adjacent modern bridge which gives a good view of the cast-iron structure, now classified as an Ancient Monument.

Further reading

JAMES J. G. The cast-iron bridges of Thomas Wilson 1800–1810. *Proc. Newcomen Soc.*, 1979, Jan.

5. Tyringham Bridge

HEW 176

SP 857 465

During his rebuilding of Tyringham Hall Sir John Soane made a new estate road, with a bridge across the River Great Ouse, to connect with the Newport Pagnell to Northampton road. His design for the bridge was a single-span segmental arch of 60 ft which was approached through the arch of the gatehouse at the entrance to the park.

The bridge is a fine if simple structure in ashlar masonry of Rutland stone. The wing walls curve boldly outwards reminiscent of Sir John's prizewinning Royal Academy design of 1776 for a Triumphal Bridge. The building work was commenced in 1793 and it is probable that the bridge was completed during that year.

Tyringham Bridge

6. Turvey Bridge

HEW 40

SP 938 524

This bridge carries the A428 Bedford to Coventry trunk road over the River Great Ouse to the west of Turvey. The 12th century bridge takes the form of a 600 ft long masonry causeway of local stone, with eleven arches varying in span from 11 ft 3 in. to 20 ft 5 in.

At one time the bridge was owned by the Mordaunt family and, in 1562, William, Lord of the Manor, left £40 for repairs. However, from time to time there were disputes about maintenance, for the same family was prosecuted at the summer assizes in 1731 for neglecting the upkeep of the bridge. The Higgins family, who succeeded the Mordaunts, suffered the same fate in 1795.

Maintenance of the structure passed to the newly formed County Council in 1889 and widening was carried out in 1931–32, in reinforced concrete faced with limestone.

Two other ancient bridges upstream from Turvey were reconstructed in the early years of the 19th century.

Olney Bridge (SP 888 509), originally built in 1619, was replaced about 1820 by a limestone masonry bridge of five segmental arches with additional flood arches south of the river. The bridge was widened in the 1970s to give a 24 ft wide carriageway.

Turvey Bridge

Near Sherington, a bridge (SP 884 453) carries a minor road across the river. An earlier bridge of eleven small spans, probably dating from the 14th centruy, was rebuilt in 1818 by Henry Provis, at a cost of £5140. This has five masonry arches, three of 40 ft and two side spans of 20 ft. The width of the bridge was originally 18 ft but this was widened to 34 ft in 1971 using concrete deck extensions, cantilevered out from each side of the earlier structure.

Further reading

JERVOISE E. *The ancient bridges of mid and eastern England.* Architectural Press, Westminster, 1932.

7. Harrold Bridge

HEW 1883

SP 955 565

Harrold Bridge is the most complete of the ancient River Great Ouse road crossings in Bedfordshire. Built on the upstream side of the original ford, the present bridge comprises three parts. The six arches spanning the river are called the great bridge and, to the south, nine arches within the flood plain form the long bridge, which carries the road down to field level. A narrow causeway several feet above ground level, containing twenty arches, then continues for a further 200 yds.

Harrold Bridge, River Great Ouse,

Although there is evidence that a bridge existed on the site in 1136 it is unlikely that any part of the existing bridge dates from that time. The varying forms of the arches reflect past ownership, repairs and the vertical curve across the bridge. Arch 15, the most northerly, has a flat segmental arch of 15 ft span whereas arches 14 and 13, of 18 ft and 15 ft respectively, are almost semi-circular. Arch 12, with the largest span of 19 ft, has a single ring of voussoirs whereas arch 11, with a span of 16 ft 6 in is more highly decorated, with three orders of stones on the downstream side. The great bridge was widened to 15 ft in 1857 with flat segmental extensions in brickwork, springing from the cutwaters. Parking is not permitted at the great bridge but access by foot is easily gained from the long bridge.

Three other noteworthy bridges carry minor roads over the river in its course towards Bedford. At Felmersham, the bridge (SP 990 579) was built in 1818 with funds raised by the villagers. It has five segmental arches and was built in limestone by two local masons, John and Samuel Bell.

Radwell Bridge (TL 005 573) was also a local enterprise when it was built in 1766 by Thomas Morris. Originally, three arches of 13 ft and two side arches of 8 ft spanned the river but two more arches were added in the 19th century.

Oakley Bridge (TL 008 528), which dates from 1815, was probably built by the Duke of Bedford's estate. The main bridge has five segmental arches and, nearby, there is an arch of irregular shape over a mill stream.

Further reading

JERVOISE E. *The ancient bridges of mid and eastern England.* Architectural Press, Westminster, 1932, 87.

8. Bromham Bridge

HEW 41

TL 011 507

The first reference to Biddenham bridge, as it was known in its early history, was in 1224, though the builder remains unknown. A chantry chapel, dedicated to Saints Mary and Katherine, was founded at the bridge in 1295 and some of its alms and land revenues were used to maintain the bridge fabric. The chapel was suppressed in 1553 and nothing now remains.

In common with other medieval bridges over the River Great Ouse in Bedfordshire, the bridge consisted of a small number of river spans, four in this case. It was wide enough for carts but the much narrower raised causeway over the flood plain was only wide enough for pedestrians and horses. Having crossed the river by the wide bridge, carts were diverted across a field.

This arrangement was clearly unacceptable for traffic in the 19th century and, in 1813–14, the six foot wide causeway was replaced by twenty new arches 17 ft wide. These linked up with the original river spans which, however, remained at their original width of 11 ft 6 in. It was not until 1902 that these spans were widened on the downstream side to the same width as the remainder. A trunk road bypass was constructed in 1986.

No traces remain of the foot causeway but fragments of the cart bridge may be seen from the bridge in the meadow on the south side.

Further reading

JERVOISE E. *The ancient bridges of mid and eastern England.* Architectural Press, London, 1932.

Bromham Bridge

9. Bedford Town Bridge

HEW 1886

TL 051 495

There is no definitive evidence of a bridge at the site of the ancient ford across the River Great Ouse until the end of the twelfth century when Simon de Beauchamp granted to St John's Hospital a chapel dedicated to St Thomas. It is thought that the first stone bridge was built of masonry from the site of Bedford Castle after the siege of 1224. There were seven arches together with two gate-houses built on the bridge, one probably a successor to de Beauchamp's original foundation. However, they constituted a considerable constriction to traffic and were removed in the late eighteenth century.

With the growth in traffic even these means were insufficient and an Act was passed in 1803 for the rebuilding of the bridge. The work was entrusted to John Wing the younger and was to be financed from public subscription which would be repaid by tolls. The new bridge was planned to be built in the same position as the old and therefore a temporary bridge had to be built alongside before the original could be demolished. Construction was completed in 1813 at a cost of £15 137.

There are five arches diminishing in span from the centre. The main structural arching, together with parapets and string courses, is in Portland stone with the

Above: Bedford Town Bridge

infill spandrels in Rutland stone. Widening was mooted a century later in 1913 but was not carried out until 1938. The widening, on the upstream side, from 30 ft to 54 ft was carried out by rebuilding the original parapets and facing stonework on to new concrete arching.

Further reading

JERVOISE E. *The ancient bridges of mid and eastern England.* Architectural Press, London, 1932, 90–92.

10. Cardington Airship Hangars

HEW 56

TL 081 468

The great sheds are located in the flat valley of the River Great Ouse, 2½ miles south of Bedford. Both sheds were built by A. J. Main & Co. in 1916–17 as steel portal frames with pin joints at the crown and in the side walls. The western shed stands in its original position but the eastern shed, first erected at Pulham in Norfolk, was re-erected on its present site in 1927.

When first built, the sheds were 700 ft long and 180 ft wide. Subsequently the length was increased by four bays to 812 ft and the height raised by 35 ft to 174 ft 6 in., the foundations being enlarged at the same time. This additional work was carried out for the Air Ministry by Cleveland Bridge Co. A total of 3700 tons of steel was used in the framework of each shed. Originally, the sheds had capstan operated doors at both ends but they now have a pair of powered doors on transverse tracks at one end only.

The hangars were used to house the new generation of British airships, developed towards the end of the 1920s, but this project was abandoned following the tragic loss of Airship R101 in 1930. The sheds are now used to provide covered space for scientific tests. Although the hangars have a recognisable silhouette from any direction their bulk is probaby best appreciated from the A600 Hitchin to Bedford road.

Further reading

LARMUTH L. Airship sheds and their erection. *Proc. Instn Civ. Engrs*, 1921, 169–225.

GIBBS A. R. Reconstruction of Cardington airship shed. *Journal IRES*, Dec. 1926.

11. Smeaton's Bridge, Cardington

HEW 1918

TL 083 484

Above: Smeaton's Bridge, Cardington

The Whitbread family were major landowners in the Cardington area from the 18th century onwards and continued to acquire land and property as this became available. Samuel Whitbread, during a particular period of expansion, was instrumental in making considerable estate improvements of which the bridge was one.

In 1778 he engaged John Smeaton to design a new bridge over a tributary of the River Great Ouse to carry the road from Cardington Village north to meet the main Bedford to Sandy road. The bridge is generally in brickwork with five segmental arches 6 ft in span and a width between parapets of 21 ft 8 in. There are triangular cutwaters at both ends, together with some unusual brickwork. The keystones of the three centre arches are marked respectively, 'J Smeaton Eng', 'SW 1778' and 'S Green Surv'.

Source: Wood J. *Bedfordshire Parish Surveys No. 3, Cardington and Eastcotts. Bedfordshire Planning Dept.*

12. Great Barford Bridge

HEW 1887

TL 134 516

Parts of this impressive bridge across the River Great Ouse are over 500 years old, built with funds left for this purpose in the will of Sir Gerard Braybrooke in 1429. A century later John Leland records the presence of a great stone bridge of eight arches. The structure spanned the river at the time but further arches were added in the 17th century to assist navigation. The presence of extra long piers suggest that one additional arch was constructed on the Barford side and five arches on the Blunham side, the year 1704 being recorded on a datestone.

Three more arches were added later and in 1818 the bridge was widened by 4 ft using the expedient of laying timbers across the cutwaters on the west side. These timbers were replaced by decorative brickwork in 1874.

Of more recent origin, nearby Tempsford Bridge (TL 162 545) was built in 1820 by John Wing to replace a timber bridge of 1735. The replacement bridge has three arches across the river with additional flood spans. Yorkshire gritstone was used for the segmental arches, cutwaters, string courses and copings whereas sandstone was used for the spandrels and parapets. The high quality of workmanship can be appreciated despite the weathering of the sandstone. The bridge carried all the traffic on the A1 road between Sandy and Huntingdon until 1961, when the southbound traffic was diverted over a new bridge.

Further reading

JERVOISE E. *The ancient bridges of mid and eastern England*. Architectural Press, Westminster, 1932, 97–99.

Great Barford Bridge

13. River Ivel Bridges: Blunham Bridge and Girtford Bridge

Blunham Bridge:
HEW 1864
TL 155 519

Girtford Bridge:
HEW 1892
TL 163 490

Blunham Bridge connects the village of Blunham with the A1 road between Sandy and Tempsford and crosses the River Ivel a mile south of its confluence with the River Great Ouse. The earliest reference to the bridge on a map is 1719 and although the style is medieval, it probably dates from the 17th century. The bridge has five stone arches, roughly semi-circular in shape, with spans between 6 ft 6 in. and 10 ft 8 in. The cutwaters, on the upstream side only, were originally carried up to the parapet to form pedestrian refuges, but these are now capped off.

Girtford Bridge remains largely unaltered since its construction in 1782 by John Wing. It has three semi-elliptical stone arches, the centre span of 29 ft and the side spans of 21 ft. The slender piers have cutwaters on both ends. A steel footbridge was erected on the south side in 1947.

Further reading

McKeague P. *Bedfordshire Magazine*, 168, 333–6.

Above: Blunham Bridge

14. Sutton Packhorse Bridge

HEW 1835

TL 221 474

Sutton High Street crosses the Potton Brook by means of a shallow ford. The bridge stands on the north side of the ford and consists of two pointed arches in sandstone masonry each of 9 ft 10 in. span with a 4 ft 7 in. centre pier. There are two orders of voussoirs to each arch but at some time the upper order over the crown has been cut away, thereby lowering the deck. Early in the life of the bridge the approaches were much shorter and the bridge must have been distinctly hump-backed in its original form.

The overall length of the bridge, including ramps, is approximately 131 ft, all in masonry except for three buttresses in brickwork supporting the south-east wing-walls. Whereas the approach ramps are over 11 ft wide the width narrows on the bridge to only 6 ft 3 in. between the parapets. A triangular cutwater on the upstream side of the pier, which is carried up to the parapet, provides a central refuge.

The earliest documentary evidence occurs in the will of Thomas Loffe of 1504. However, radiocarbon dating of foundation timbers, uncovered during major repairs in 1986, suggests 13th century construction, a date which would be in keeping with the style of the arches. The high

Above: Sutton Packhorse Bridge

quality of the work would suggest that construction was the responsibility of the Lord of the Manor of Sutton.

Further reading

McKeague P. Sutton Packhorse Bridge. *Bedfordshire Archaeology*, 18, 1988.

15. Wadesmill Bridge, Ware

HEW 1927

TL 359 175

As early as 1621 there was a wooden bridge over the River Rib at this site but in 1674–5 the bridge was rebuilt in brickwork with five arches. At the Hertfordshire Session in 1768 the County was required to repair the bridge which was in a dangerous and ruinous condition. Richard Norris, a bricklayer of Castleyard in London was ordered to make a survey and submit a report.

The bridge presumably deteriorated again, for in 1824 a Mr Robson submitted plans and specifications for a two-span brick arch bridge for the Cheshunt Trust. Brough and Smith of London carried out the reconstruction in 1825, accompanied by a road realignment.

The bridge design was unusual in that the pier consists of six tapered columns in Cornish granite with bedstones and springing capping stones of the same material. Stone is also provided for the abutment springings, the string course and the wingwall copings. The arches are segmental in shape with a span of 25 ft and the bridge is 31 ft wide. Following vehicle damage the decorative lattice iron parapet was replaced by a more utilitarian steel type with vertical infill bars.

Further reading

Jervoise E. *The ancient bridges of mid and eastern England*. The Architectural Press, London, 1932.

16. Midland Railway

HEW 1924

TQ 38 to SK 33

Following its formation in 1843 from the amalgamation of three Derby centred railway companies, the Midland Railway sought a way to achieve a connection to London in order to exploit the growing volume of traffic from the midlands, particularly in coal. The aspiration of a direct line to London was only to be achieved by stages, spread over the next 24 years. In view of the cost of a complete new route from Leicester to the capital, the Midland at first relied on its connection with the London and Birmingham (L & B) Railway at Rugby to handle its London trains.

The passage over the L & B railway proved to be rather restrictive and, in 1846, the Midland explored a scheme to extend from Leicester to Bedford. Powers were obtained in 1847 but, owing to financial constraints at the end of the railway mania period, the Act was allowed to lapse. However, in 1851, with a more buoyant and confident economy, the scheme was revived with the addition of an extension to Hitchin to link up with the Great Northern Railway (GNR). Work started in Feburary 1854, with Thomas Brassey as the contractor and the line was opened in May 1857.

The route crossed several hilly areas involving substantial gradients at Kibworth and Desborough. The most severe gradient applied to the long climb of 1 in 120 to Sharnbrook Summit. Later, with the growth of heavy freight traffic, it became necessary to obtain a less demanding gradient of 1 in 200 by the building of a diversion for goods trains to bypass the summit. This required a tunnel 1860 yd long. Major works on the Bedford route included bridging the River Nene at Wellingborough, six large bridges over the River Great Ouse between Sharnbrook and Bedford and another river crossing at Bedford.

By 1862 the Midland was experiencing delays to its

Midland Railway, Derby

London traffic over the GNR and it resolved to promote its own line into London from Bedford. The Midland Railway London Extension Bill was passed in June 1863. Charles Liddell was appointed engineer and the contract was awarded to Brassey and Ballard. Work started in April 1865 and was completed in August 1867. The route, passing through Luton and St Albans, required heavy earthworks, mainly in London clay, tunnels at Ampthill, Elstree and one of 1926 yd at Hampstead.

The Midland was noted for the quality of its stations, illustrated by fine examples at Kettering and Loughborough, with hipped ridge and furrow awnings supported on slender columns with filigree brackets. But its showpiece was St Pancras station, opened in 1868. The hotel block, which forms the front of the station, was designed by Sir George Gilbert Scott in an exuberant Gothic style and it remains the most ostentatious of all the London termini. No less dramatic was the trainshed, (HEW 237) engineered by W. H. Barlow, with its great sweeping roof, 240 ft wide, the largest single-span roof in the world at the time. From its rather hesitant beginnings the Midland was to reach far and wide until its incorporation in the London Midland and Scottish Railway.

Further reading

BARNES E. G. *The rise of the Midland railway, 1844–1874*. George Allen and Unwin, London, 1966.

17. Great Northern Railway

HEW 1921

TQ 38 to SE 50

The construction of a direct rail link from York to London came after the frenetic rush to build the first trunk routes, such as the London to Birmingham and London to Bristol railways. By the mid 1840s, York already had a connection to London via a devious route over several lines in the Midlands. The proposal to build a direct line through the eastern counties was vehemently opposed by George Hudson, the railway mogul of York. Despite this formidable opponent, the sponsors of the direct line were able to achieve success and the line became a key element in the national rail network.

The choice of routes between Doncaster and Peterborough lay between a line through Lincoln and a route connecting the towns of Retford, Newark and Grantham. In the event, the London and York Railway was authorised in 1846 to build both routes, the largest single railway scheme which had then been approved. The Great Northern Railway, formed from an amalgamation of the London and York and the Direct Northern Railway, appointed William Cubitt as the engineer, assisted by his son Joseph.

By 1848, Peterborough was connected to Lincoln and in 1852 the more direct route via Retford was completed. A ruling gradient of 1 in 200 presented few obstacles, although two tunnels were needed at Stoke and Peascliffe, to the south and north of Grantham respectively.

South of Peterborough, the engineer, and his contractor Thomas Brassey, faced a more exacting task. The River Nene at Peterborough was crossed with a large cast-iron bridge (HEW 93), which still remains in use. A few miles further south the line had to cross the marshy area of Whittlesea Mere and the solution adopted was to sink a series of brushwood mats to support the embankment. This was an expedient used at Chat Moss on the Liverpool and Manchester Railway (HEW 952).

To cross the valley of the River Mimram, north of Welwyn, Joseph Cubitt designed an imposing brick viaduct (HEW 159), rising to about 100 ft, with 40 arches. Passing the route through the line of hills in this area required a sequence of tunnels, at Welwyn, Potters Bar and Hadley Wood (HEW 1922) and at Barnet and Wood Green. The incline approach to Kings Cross also needed tunnels, Copenhagen and Gasworks, to pass under the Regent's Canal and Caledonian Road. The line was opened to a temporary terminus at Maiden Lane, between these two tunnels, in 1850, and into the new Kings Cross station two years later.

The prestige mark of the Great Northern was Kings Cross station, opened in 1852. Designed by Lewis Cubitt, a nephew of William Cubitt, the twin arched facade to the train shed is regarded as one of the finest examples of restrained and functional railway architecture.

Further reading

GRINLING C. H. *History of the Great Northern railway, 1845–1895.* Methuen, London, 1898.

18. Deards End Lane Bridge, Knebworth

HEW 279

TL 247 208

The bridge was designed by Joseph Cubitt in 1849 to carry a minor road over the Great Northern Railway, now part of the East Coast main line.

The site, just north of Knebworth Station, is in a deep cutting. Anticipating excellent support from the hard chalk, Cubitt placed the abutments halfway up the cutting slopes and designed a segmental arch of 96 ft span with a rise of 20 ft 9 in. Outwardly there is nothing unusual about the bridge but an arch of this size plainly demanded as much saving in weight as possible, commensurate with great strength.

These objectives were achieved by building the arch soffit in five rings of brickwork which then supported five longitudual ribs, each only 2 ft 3 in. thick but 9 ft deep at the haunches, tapering to 4 ft 6 in. deep at the crown. The ribs are topped by 6 in. thick York stone slabs which carry the road surfacing.

The wing walls have stepped foundations following the slope of the cutting and both these and the abutments are cellular in form with slender internal piers surmounted by semi-circular arch rings.

19. Welwyn Viaduct

HEW 159

TL 248 147 to TL 246 151

Within a distance of fifteen miles from Hadley Wood to Knebworth the Great Northern Railway was faced with constructing a succession of tunnels, cuttings, embankments and bridges in its route over the Eastern Chilterns and the valleys of the Rivers Lea and Mimram. The crossing of the Lea was by means of a bridge and embankment but the deeper valley of the Mimram called for a viaduct.

This imposing brick structure is 520 yd long and consists of 40 arches, semi-circular in shape, each 30 ft in span. It carries two tracks 98 ft above the valley floor and the width between the parapets is 26 ft. Construction of the arches is somewhat similar to the single span overbridge at Deards End Lane, Knebworth (HEW 279) in that the four rings of brickwork to each arch are strengthened by three internal longitudinal ribs.

Both abutments and piers are cellular and the piers taper inwards in both directions at a rake of 1 in 40. Approximately 13 million bricks were required for the structure, made from local clay.

Joseph Cubitt was the designer and the viaduct work was included in the 75 miles of Thomas Brassey's contract between Peterborough and London. The contract was started in 1846 and completed in August 1850.

The viaduct is best viewed from the B1000, Welwyn to Hertford road. There is a convenient car park at the junction with the Digswell Road from which it is possible to appreciate this impressive structure at close quarters.

20. Potters Bar and Hadley Wood Tunnels: Potters Bar, Hadley Wood North and Hadley Wood South

Potters Bar HEW 1922 TL 256 004–TQ 261 994

Hadley Wood North TQ 262 987–TQ 262 985

Hadley Wood South TQ 262 981–TQ 262 978

As part of the Great Northern Railway, three tunnels were required through the hilly ground between Barnet and Potters Bar and each was constructed to take two tracks. Anticipated additional traffic, following modernisation after the second World War, made it imperative that additional tunnels to carry a further two tracks were constructed at all three sites. They formed part of a quadrupling scheme about two and half miles long between Greenwood and Potters Bar.

Parliamentary powers were obtained in 1954 and work commenced in the following year to a design by Sir William Halcrow and Partners for the Chief Civil Engineer, Eastern Region British Railways. In view of the prevailing scarcity of both skilled bricklayers and engineering bricks, the method adopted to line the tunnels consisted of interlocking precast concrete units. Specially constructed shields 31 ft in diameter, each with four decks, were employed, the London clay being excavated by hand using pneumatic clay spades. Spoil was removed to tip by conveyor and trains of wagons on Decauville track.

Lining segments were erected by hydraulic arms

mounted on the shield and pressed into the clay. Jacks were then placed in a gap left between the upper arch ring and lower invert and pressed further against the ground. Dry concrete was placed in the gaps to complete the rings. When required, the shield was moved forward by hydraulic rams bearing against the completed rings.

The inside diameter of all three tunnels is 26 ft 6 in. with the crown 19 ft 6 in. above rail level. The Potters Bar Tunnel, being 1214 yd long, is provided with a 12 ft diameter ventilation shaft, but none were needed for either Hadley North, 232 yd long or Hadley South, 384 yd long.

The work took nearly four years to complete and cost £2.5 million. The Contractor was Charles Brand & Co.

Further reading

TERRIS A. K. and MORGAN H. D. New tunnels near Potters Bar in the Eastern Region of British Railways. *Proc. Instn Civ. Engrs. 1961,18,289–304*

21. Ponsbourne Tunnel

HEW 1923

TL 309 052 to TL 315 077

The loop line between Wood Green, Hertford and Stevenage was completed in 1918 by the Great Northern Railway. It served Enfield and Hertford but also served as a valuable relief line until the tunnels at Potters Bar and Hadley Wood were doubled nearly forty years later.

The whole of the one and a half miles of the tunnel, the longest on the Great Northern Railway is constructed within London clay overlying chalk. It was built by traditional methods, brick lined throughout in a horseshoe shape cross-section, with an invert. In order to take double tracks the span is 27 ft with a height of 20 ft 9 in.

Further reading

BLOWER A. *British Railways Tunnels.* Ian Allan, London, 1964, 42.

22. Gadebridge Bridge

HEW 1802

TL 052 080

Spanning the River Gade to the north of central Hemel Hempstead, this graceful bridge has a pleasant setting in the midst of parkland. It is easily accessible from the Leighton Buzzard road A4146, with a car park at the site.

Originally constructed to give access for carriages to the High Street from Gadebridge House, now demolished, the 34 ft 6 in. span now carries a broad footpath only. Four cast-iron arch ribs, each in two parts with a transverse stiffener, link the halves at midspan. The upper and lower members of the main arch ribs are connected by integral hoops to form spandrels. The parapets have an elaborate pattern simulating bamboos, which splay out at the terminal pilasters.

A masonry pilaster over the crown bears the inscription 'J Cranstone' and the springing plate on the south-west abutment bears the inscription 'Barwell and Company Northants'. At the time of construction, which was probably about 1840, both firms undertook casting work. Whereas Barwell and Co. were certainly capable of, and had indeed provided, larger cast-iron elements, there is no comparable evidence that Cranstone could do the same. It is likely therefore that Cranstone acted as agent for the supply of the main units.

Further reading

COX A. *Notes on Edward Harrison Barwell, Ironfounder, Northampton.* Bedfordshire County Council, 1981.

BRANCH JOHNSON W. *The industrial archaeology of Hertfordshire.* David and Charles, Newton Abbot, 1970, 114, 125.

23. Tring Cutting

HEW 48

SP 951 122 to SP 931 152

The London and Birmingham Railway, now part of the West Coast route, runs through a narrow Chiltern valley between Hemel Hempstead and Berkhamstead. At Tring however, the valley widens and the railway runs in cutting from Tring Station for a distance of 2 1/2 miles before crossing the Vale of Aylesbury. The line was designed for easy curves and gradients, making heavy earthworks inevitable. This cutting was the longest on the first major trunk railways in the country.

In order to take a double track the cutting was constructed to a width of 33 ft at rail level with a longitudinal falling gradient to the north of 1 in 330. The side slopes were almost 1 to 1 and whereas the average depth is 40 ft, there is a maximum depth of nearly 60 ft.

Excavation at that time was a manual job except where blasting was required, the volume of excavation being 1½ million cu. yd. Barrow runs were spaced out along the cutting slopes every 75 ft and spoil was hauled up the

Tring cutting. J. C. Bourne

runs to platforms at the top of the cutting, where the material was tipped into haul wagons for spreading. The method of excavation is excellently portrayed in the well known lithograph by J C Bourne.

In the late 1850s the line was quadrupled, but as far as is known no further land was purchased to extend the boundary fences. The additional tracks were evidently accommodated by steepening the side slopes still further. The three bridges crossing the cutting were reconstructed during the widening but remained as three span arches with new centre spans of 68 ft.

Further reading

ROSCOE T. *London and Birmingham Railway*. Charles Tilt, London, 1838.

24. Pitstone Green Windmill

HEW 1768

SP 945 157

Pitstone Mill is situated part way up the scarp of the Chiltern Hills above the Vale of Aylesbury. It stands on a small plateau between the lower and upper Icknield Ways, with the ground rising behind for another 300 ft.

The timber clad buck carries four 27 ft common sails and stands on a roundhouse 22 ft in diameter. There is no fantail and the mill is luffed by a tail pole. The oldest date found on the windmill is 1627 carved on a lower side

Pitstone Green Windmill

timber; assuming authenticity, this is the earliest date on any existing mill.

In 1840 the mill was owned by the Grand Junction Canal company who sold it to Francis Beesley in 1842. He worked it for 32 years and then sold the mill to Lord Brownlow, owner of the Ashridge Estate, who in turn let the mill to the Hawkins family, tenants of a nearby farm.

Major repairs were carried out in 1895 but in 1902 the mill was badly damaged. A sudden squall caught the sails before they could be turned to the wind and the windshaft become unseated from the tail bearing. As the windshaft rotated the sails caught the roundhouse.

Repairs were uneconomic and the mill fell into disrepair. In 1937 the Hawkins family, who were now the owners, gave it to the National Trust. Some work was done as a holding operation but it was not until 1963, with the formation of the Pitstone Mill reconstruction committee, that the mill was restored to its present condition.

Further reading

BROWN R. J. *Windmills of England*. Robert Hale, London, 1976, 48–49

25. Tring Reservoirs

HEW 1767

SP 92 13

The purpose of the Grand Junction Canal was to connect London with the great manufacturing centre of Birmingham. By 1790 the rival Oxford Canal had been constructed but it followed a circuitous route with over one hundred locks and the sometimes hazardous navigation weirs on the Thames.

The Grand Junction chose a more easterly route, but by so doing had to negotiate the Chiltern Hills. Long flights of locks were required and hence large quantities of water. It was intended that the natural drainage from the hills would be picked up and fed to the main line by the Wendover feeder, nearly seven miles long.

However, once the canal was opened it became clear that further water supplies would be required. Matters were not helped by severe leakage from the feeder. In the forty years following the opening of the canal there was continual development of the reservoirs, headings and pumping stations.

The first reservoir, Wilstone No. 1, was constructed in 1802 followed by Marsworth in 1806. Tringford high level and Startopsend were added next, being necessary to water the Aylesbury Branch. Wilstone was enlarged in 1811 and again in 1817 and further reservoirs were added at the Wilstone site in 1836 (No. 2) and 1839 (No. 3).

Further reading

FAULKNER A. H. *The Grand Junction Canal.* David and Charles, Newton Abbot, 1972, 132–144 and 205-208

RICHARDSON A. Water Supplies to Tring Summit. *Journal of the Railway and Canal Historical Society*, April 1969.

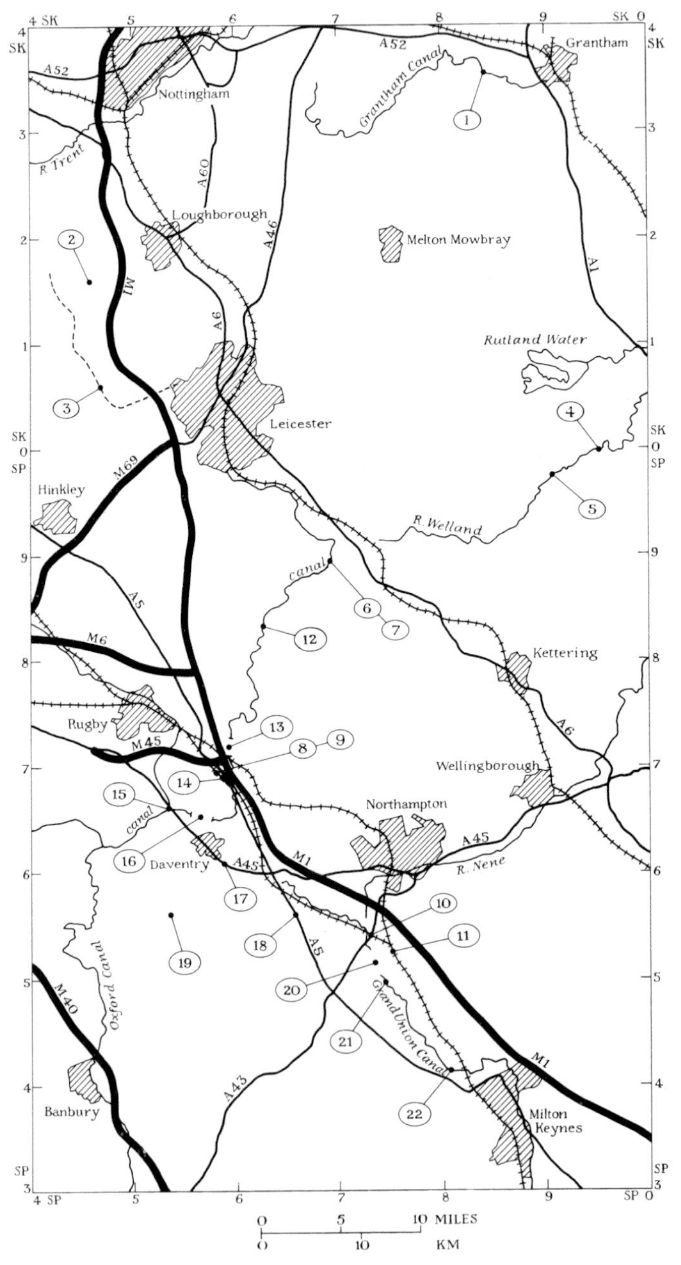

1 Grantham Canal
2 Blackbrook Dam
3 Leicester and Swannington Railway
4 Wakerley Bridge
5 Harringworth Viaduct
6 Foxton Locks
7 Foxton Inclined Plane
8 London and Birmingham Railway
9 Kilsby Tunnel
10 Blisworth Railway Bridge
11 Road Cutting
12 Grand Union Canal
13 Crick Tunnel
14 Watford Locks
15 Braunston Towpath Bridges
16 Braunston Tunnel
17 Holyhead Road
18 Watling Street Pavement
19 Charwelton Pack-horse Bridge
20 Blisworth Tunnel
21 Barge Weighing Machine, Stoke Bruerne
22 Cosgrove Aqueduct

7. Leicestershire and Northamptonshire

These two counties have many similarities. The largest town of each one is the county town, which stands in a central position, with smaller manufacturing and market towns around them; Loughborough and Melton Mowbray north of Leicester, Kettering and Wellingborough east of Northampton, and Market Harborough between them.

Running south-west to north-east, passing between Northampton and Leicester, is a ridge of high ground, part of the Jurassic (oolitic) limestone belt that crosses the country from the Cotswolds to the Lincoln Edge. It is a prolific source of building stone, which characterises most of its villages. The major river of the area is the Nene, which rises in this ridge and flows east and then north-east through Northampton and Wellingborough. A lower ridge running east separates the Nene valley from that of its more southern neighbour, the River Ouse. The River Soar rises west of the high ridge and flows northwards through Leicester and Loughborough to join the Trent. Another notable geological feature of Leicestershire is the Charnwood Forest area north-west of Leicester, an outcrop of ancient, igneous rocks, notably granite, which is extensively quarried.

Over the past two thousand years the area has been crossed by every mode of transport. The Romans were the first to traverse it, Watling Street running diagonally from south-east to north-west, now forming the A5 trunk road and, for part of the way, Thomas Telford's Holyhead Road. The early canal builders were also active in this area. In the west, the Oxford Canal runs south from the Midlands towns to the Thames whereas the later Grand Juction Canal which eventually formed a major part of the Grand Union network, follows much of the Roman route. In the north of the area, the waterway route between London and the River Trent was completed by the opening in 1810 of the (old) Grand Union Canal with its famous staircase of locks and inclined plane at Foxton, which finally linked the Loughborough and Leicester Navigations and the Leicester and Northamptonshire Union Canal with the Grand Junction Canal.

Between Northampton and Rugby there is a significant lowering of the ridge of high ground separating the two towns (the Watford Gap) and it is interesting to observe how, following the example of the canal pioneers, all the later transport modes including the London and Birmingham Railway and the M1 Motorway have taken advantage of this natural feature. For a distance of several miles only a few yards separates the routes of the canal, railway and motorway. However, despite the existence of this natural crossing point major engineering works were still required for both the canal and the railway in this area and, due to the treacherous nature of the ground, great difficulties were encounted during construction. The stories of the building of the Blisworth, Braunston and Crick canal tunnels and the great railway cutting at Roade and tunnel at Kilsby will be found within this section.

1. Grantham Canal

HEW 1850

SK 583 383 to SK 908 353

This is the longest of ten canals engineered by William Jessop in south Nottinghamshire and the adjoining counties. It was constructed between 1793 and 1797 as a 33 mile waterway connection, with 18 locks, between Grantham and the River Trent at West Bridgford, near Nottingham.

It was the first canal to draw its water supply almost entirely from reservoirs, of which there are two, at Knipton, SK 81 30, and Denton, SK 87 33. Knipton reservoir covers an area of 50 acres. The dam, the largest in the country when it was built, is 240 yd long, in a steep-sided valley on private land owned by the Belvoir Castle estate. The other reservoir, near Denton, Lincolnshire, is almost surrounded by a 1500 yd curved embankment along which runs a public footpath.

The canal is no longer navigable (except by canoes) but the towpath is walkable throughout its length. It was abandoned in 1936 but a canal society has taken on the task of clearing the towpath and the canal bed, and hopes to restore the canal for navigation. This would involve making a new entrance from the Trent and a new terminal basin at Grantham, both ends of the canal having been obliterated by road works.

In the National Railway Museum at York there is a wagon that used to run on a horse-drawn tramway between the Grantham canal and Belvoir Castle, which operated between 1815 and 1918. The wagon is described as 'the oldest vehicle with flanged wheels to be preserved in Britain'.

Further reading

HADFIELD C. and SKEMPTON A. W. *William Jessop, Engineer*. David and Charles, Newton Abbot, 1979.

HADFIELD C. *Canals of the east Midlands*. David and Charles, Newton Abbot, 1981.

2. Blackbrook Dam

HEW 1825

SK 457 158

The first dam in the valley of the Black Brook, 5 miles west of Loughborough, was an earth dam built under the direction of William Jessop in 1797 to provide a water supply for the newly opened Charnwood Forest Canal (strictly called the Forest Line of the Leicester Navigation) which connected the coal pits around Coleorton to Nanpantan, the remaining 2½ miles to the canal at Loughborough being traversed by a descending tramway. This was the first tramway on which iron edge rails as distinct from L-plates were used. The Blackbrook reservoir dam was breached by a flood in 1799 and, although it was repaired, the mishap heralded the end of the canal's short life.

A second dam, which still stands, was built in the same valley between 1901 and 1906 to create a reservoir to augment the public water supply of Loughborough. It is of mass concrete, with brick facing to the upstream side, 525 ft long and 65 ft high from valley floor to overflow sill. The thickness of the dam is 65 ft at the bottom and 15 ft at the crest, with a roadway carried along the top on six masonry arches, each of 25 ft span. The reservoir has a capacity of 506 million gallons and a surface area of 84 acres. The engineer of the scheme was George Hodson, a former surveyor to the Loughborough Board of Health, and the construction was carried out by Loughborough Corporation's own labour force. The dam and reservoir are now owned and operated by Severn-Trent Water.

On 11 February 1957 an earthquake shock was felt over much of the East Midlands; its magnitude was over 5 on the Richter scale and its epicentre was only 4 miles north of the Blackbrook Dam. A special inspection of the dam was immediately undertaken and it was seen that all the coping stones had been lifted, daylight being visible under many of them; the whole dam appeared to have been lifted bodily and then fallen back, but the overspill sill remained level. A few cracks appeared in the dam, but they later closed up, and no increase was seen in the regular slight leakage through the dam, which has continued in service — a tribute to Hodson's design.

3. Leicester and Swannington Railway

HEW 1745

SK 579 045 to SK 417 167

The Leicester and Swannington Railway was one of the first standard-gauge lines to be built in the Midlands. It was promoted by the coal-owners of west Leicestershire

Leicester and Swannington Railway, Soar Lane Bridge, Leicester

who had observed the success of the Stockton and Darlington Railway (1825) and had approached George Stephenson for his advice. Stephenson advised on the route (and on the gauge) and his son Robert became the engineer for the work.

The line was 16 miles long, single track, running north-west from West Bridge, Leicester, to Swannington in the middle of the coal-mining area. Its route included a 1 mile tunnel at Glenfield, an ascending self-acting incline at Bagworth (1 in 29) and a descending cable-hauled incline at Swannington (1 in 17). The first 11 miles to Bagworth were opened in 1832, and the remainder in 1833. The Midland Railway took over the company in 1846 and constructed a new double-track line from Leicester to Burton-on-Trent, still in use as a British Rail goods line. This followed much of the route of the Leicester and Swannington Railway, but bypassed the tunnel and the two inclines. The other parts of the old route remained in use until 1966 except for the Swannington incline, which closed in 1947. The winding-engine from that incline, an early horizontal engine by the Horseley Iron and Coal Co., is now in the National Railway Museum, York.

The west end of the Glenfield tunnel is still visible, though difficult of access. The east end has disappeared in an infilled cutting and so have some of the tunnel's brick vent shafts and the two inclines.

4. Wakerley Bridge

HEW 59

SP 955 997

Above: Wakerley Bridge

An excellent example of a 14th century stone arch bridge, and typical of many bridges of its period, Wakerley Bridge carries a minor road between Wakerley and Barrowden over the River Welland. It has five arches, the original arches on the downstream side being pointed with chamfered arch rings. The bridge was widened by 2 ft on the upstream side in 1793, the new arches being segmental. The spans of the arches vary from 9 ft 10 in. to 11 ft 4 in. with stone piers about 6 ft wide. The roadway over the bridge is 12 ft wide between stone parapet walls which are 15 in. thick and up to 3 ft 8 in. high. The total length of the bridge is 81 ft.

There are cutwaters on both sides of the bridge and above the keystone of the second arch from the north end of the bridge is a carved head.

Further reading

JERVOISE. E. *The ancient bridges of mid and eastern England.* Architectural Press, London, 1932, 63–4.

5. Harringworth Viaduct

HEW 417

SP 913 975

In the early 1870s, the Great Northern Railway and the London and North Western Railway proposed a new line from the Nottingham area to Market Harborough. This led to a rival proposal by the Midland Railway to open up a new route from Nottingham to Kettering via Melton Mowbray. This new route was also designed to relieve the main Midland line to the south via Leicester.

The route involved the construction of two new lines, Nottingham to Melton Mowbray and Manton (south of Oakham) to Glendon Junction (north of Kettering). The Manton-Glendon Junction line was fully opened to traffic, including through trains from St Pancras to Nottingham, on 1 March 1880.

In order to carry the line across the wide valley of the River Welland north of Corby, a long viaduct was required, the railway passing through tunnels to the north and south of it. This is the longest viaduct on the British Railway system excluding those carrying the London suburban lines.

The viaduct lies just north of Harringworth Station (closed in 1948) and is of brick construction, 3825 ft long carrying a double line of railway. There are 82 semicircular arches, each of 40 ft span, made up from six courses of brickwork. The maximum height of the structure is 60 ft.

The designer was John Underwood.

Further reading

Leleux R. *A regional history of the railways of Great Britain.* Volume IX, *The east Midlands.* David and Charles, Newton Abbot, 1976, 79–81.

Harringworth Viaduct

Foxton Locks, detail of lock chamber. Paddle gear to and from side ponds on left

6. Foxton Locks

HEW 46

SP 691 895

In 1793 the Leicester and Northamptonshire Union Canal Company received its Act of Parliament to construct a canal from Leicester to Braunston where it would join the Grand Junction Canal and thus give access to London and Birmingham. The canal was built southwards from Leicester and by 1797 had reached Gumley Debdale Wharf near Foxton. Here construction ceased due largely to lack of funds and, apart from an extension of the canal to nearby Market Harborough (reached in 1809), the original company carried out no further construction work. In 1799, James Barnes investigated the possibility of continuing the canal south to Braunston and later a second survey was carried out by Thomas Telford. However it was not until 1810 that the 'old' Grand Union was granted an Act to build an extension to Braunston via Watford. Benjamin Bevan engineered the route, adopting the line originally surveyed by Barnes.

The new canal branched off the old Leicester and Northamptonshire Union Canal at Foxton and was immediately faced with a steep ascent of 75 ft to reach the summit level. To overcome this difference in elevation a remarkable system of ten locks was constructed, consisting of two staircases, each of five locks, with a short pound in between the staircases to allow boats to pass. The lock chambers are built in brickwork with stone copings and are 7 ft 9 in. wide and have an average rise of 7 ft 6 in. In staircase locks the lower gates of one chamber form the upper gates of the next lower chamber and at Foxton the gates are double mitred gates throughout except for the topmost gates of each staircase which are single leaf gates. Side ponds are provided on the east side of the staircase and each lock chamber has two ground paddles set in the lock wall to control the flow of water between the lock chamber and the side ponds.

The locks were opened in 1814 and were used until 1900 when the adjacent inclined plane lift (HEW 45) was brought into use to reduce the delays caused to traffic in negotiating the staircase locks. The lock chambers were reconditioned and brought back into use in 1909–10 when the inclined plane was abandoned. The double staircase of locks at Foxton is probably the greatest example of the use of staircase locks in Britain.

Further reading

HADFIELD C. *The canals of eastern England.* David and Charles, Newton Abbot, 1981, 97–107.

7. Foxton Inclined Plane

HEW 45

SP 692 895

Towards the end of the 19th century, the great flight of locks at Foxton (HEW 46) had fallen into such a state of disrepair that it formed a severe bottleneck to traffic on the canal. The Grand Junction, which purchased the canal in 1894, decided to bypass the locks by building an inclined plane boat lift which would greatly speed up the passage of boats and also, incidentally, save water. The lift was designed by the Company's Engineer, Gordon Thomas and built by J. & H. Gwynne.

The lift was formed from two large steel tanks each 80 ft long, 15 ft wide and 5 ft deep, which ran on rails up and down a slope of 1 in 4, over 300 ft long. The tanks were mounted with their long axes parallel to the slope of the plane and were connected together by a single balancing wire rope attached at the centre of each tank and by two hauling wires connected to the ends of the tanks. The hauling ropes passed round a single winding drum ar-

Foxton Inclined Plane with the engine house under restoration

ranged so that as one tank was raised the other was lowered and the system was constantly in balance. By means of watertight gates at both ends the tanks were kept full of water at all times. At the top of the plane the tank was forced by hydraulic rams against the end of the canal which was also closed off by a gate. Both the tank and canal gates were then raised, again by hydraulic power, and boats passed into and out of the tank. At the bottom of the plane the tank was immersed in the water in the canal, the tank gate being raised when the water levels in the tank and the canal were equalised.

The lift system was operated by a steam engine of about 25 hp supplied with steam from one of two boilers. The engine powering the lift also supplied water under pressure for the operation of the hydraulic rams.

The lift, which opened in July 1900, was operated successfully for ten years but closed in 1910 as it was found that the traffic was insufficient to justify its continued use. The locks were repaired and brought back into use. Little now remains of the lift but the concrete foundations of the rails on the slope may be seen and the engine house is undergoing restoration. The lift site is a designated Ancient Monument.

Further reading

HADFIELD C. *The canals of the east Midlands*. David and Charles, Newton Abbot, 1981, 225–227.

8. London and Birmingham Railway

HEW 1092

TQ 294 828 to SP 078 871

Two routes were originally proposed for a railway from London to Birmingham, one by Sir John Rennie via Banbury and Oxford and one by Francis Giles via Coventry. George Stephenson was engaged by the Coventry route promoters and gave a favourable opinion of that route. An Act for the construction of the railway was obtained in 1833 and Robert Stephenson was appointed engineer. The route of the railway, via Watford, Bletchley, Milton Keynes and Rugby is still a major part of the West Coast route to Scotland, although trains not calling at Birmingham now leave the London and Birmingham at Rugby and travel via the Trent Valley line to Stafford. Originally built as a double track line, the section between Euston and Roade, near Northampton, was quadrupled between 1859 and 1882.

Construction of the 112 mile line began in 1834, a total of 30 contracts being let for lengths of line or individual structures. The London station originally proposed at Hampstead Road was replaced by the present terminus at Euston which involved trains negotiating the Camden Bank, 1 mile long at a gradient of 1 in 70. The first section, from Euston to Boxmoor was opened in July 1837. Althougth the railway reached Birmingham in April 1838, severe construction difficulties on the section east of Northampton where the line cuts through a ridge of high ground delayed the complete opening of the line until September 1838. Between April and September a road coach service bypassed the still incomplete section between Denbigh Hall (now part of Milton Keynes) and Rugby. In 1882 a new 'loop' line was opened from Roade to Rugby, passing through Northampton.

Many fine structures were built for the line including Watford Tunnel (1 mile 55 yd), Denbigh Hall Bridge over the A5 (HEW 415, p181) and the six-span Wolverton Viaduct over the River Ouse (HEW 152,p.182). A description of the great cuttings at Tring (HEW 48, p.202) and

London and Birmingham Railway. J. C. Bourne

Roade (HEW 50) and the 2426 yd tunnel at Kilsby (HEW 55), the construction of which delayed the opening of the railway, will be found elsewhere in this volume.

The terminus stations of the London and Birmingham Railway were designed by Architect Philip Hardwick in the classical style. Unfortunately Hardwick's great Doric portico at Euston was destroyed in the rebuilding of the station, but the station at Curzon Steet, Birmingham (HEW 420) with its contrasting Ionic columns has survived (see *Civil engineering heritage Wales and western England* pp 143–145).

The London and Birmingham Railway merged with the Grand Junction and Manchester and Birmingham Railways in 1846 to become the London and North Western Railway. This in turn became part of the London Midland and Scottish Railway in 1923 and of British Railways in 1948.

Further reading

ROSCOE T. *The London & Birmingham Railway. Charles Tilt, 1838.*

LELEUX, R. *A regional history of the railways of Great Britain, Volume IX, The East Midlands.* David and Charles, Newton Abbot, 1976, 12–25, 43–52.

9. Kilsby Tunnel

HEW 55

SP 578 698 to SP 565 714

Between Rugby and Northampton lies a ridge of high ground which was a formidable obstacle to the construction of both canals and railways. Within a few miles it is pierced four times, twice by canal tunnels at Braunston (HEW 39) and Crick (HEW 1729) and twice by railway tunnels. Kilsby Tunnel, the earlier of the rail tunnels was built between 1834 and 1838 during the construction of the London & Birmingham Railway. The second rail tunnel was built at Crick to carry the Northampton to Rugby line through the ridge.

Kilsby Tunnel proved to be an enormous and costly task, ultimately delaying the opening of the line. The tunnel, designed by Robert Stephenson, is 2426 yd long. At the time, it was by far the longest tunnel to be built for steam engines and Stephenson designed it to be 28 ft high to combat the expected fear that passengers would suffocate. He also included two huge ventilation shafts of 60 ft

Kilsby Tunnel.
J. C. Bourne

diameter which can be seen when travelling along roads over the line of the tunnel. A total of 18 shafts were planned to be sunk during the construction.

It was known from the earlier driving of the canal tunnels that the strata was deeply faulted with a threat of quicksands. However, trial borings along the intended line of the tunnel did not reveal any exceptional hazards in the ground conditions.

Very shortly after work began in 1834 disaster struck. Quicksands were encountered in the headings and a great volume of water was released into the workings. The contractor withdrew defeated and the railway company had to take over the conduct of the work. George Stephenson visited the site with his son Robert and they adopted the solution of pumping out the flooded workings, a task which was to be long and arduous. A further seven shafts were sunk and a total of 13 pumps were installed, delivering 2000 gallons a minute but it took 19 months of pumping before the ingress of water was brought under control. The tunnel was finally completed in June 1838.

Further reading

LELEUX R. *A regional history of the railways of Great Britain*. Vol. IX, *The east Midlands*. David and Charles, Newton Abbot, 1976, 51.

10. Blisworth Railway Bridge

HEW 418

SP 729 541

An elegant stone and brick bridge carries the London and Birmingham Railway over the A43 trunk road at Blisworth. It was designed by Robert Stephenson and has a single semicircular arch of 30 ft span which springs from vertical brick walls 17 ft 6 in. high. The interior of the arch is in brickwork and the exterior is in rusticated ashlar masonry. The bridge is 31 ft wide and carries a double line of railway. At the sides of the arch are tapering buttresses 10 ft wide.

The wing walls of the bridge are vertical and parallel to the arch, the embankment being retained by rough stone walls built at right angles to the wing walls. There is a neat dripstone moulding below the low parapet walls.

This is a fine example of the attention to detail in bridge design which was typical of the early railway engineers.

11. Roade Cutting

HEW 50

SP 758 512 to SP 745 532

In order to carry the line of the London and Birmingham Railway through high ground to the east of Blisworth Hill, Robert Stephenson designed a deep cutting between Roade and Blisworth in Northamptonshire. The cutting is about 1½ miles long with a maximum depth of about 65 ft. The cutting is spanned by several tall brick bridges.

The cutting passes through very poor ground consist-

Roade Cutting.
J. C. Bourne

ing of bands of limestone, shale and clay and in order to hold back the sides of the cutting very thick brick retaining walls were found to be necessary. Major problems were also encountered because of the ingress of large quantities of water.

The construction of the cutting was originally undertaken by William Hughes as contractor, but, largely owing to the difficulties caused by the bad ground, he was unable to complete the job and the works were taken over by the London and Birmingham Company.

In 1881–82 the cutting was widened and deepened on its east side to allow for the quadrupling of the tracks. Towards the north end of the cutting the new lines, forming the Northampton loop line, dip below the level of the Rugby line. New brick retaining walls were built on the east side of the cutting and between the two sets of tracks with an inverted arch underneath the deepened lines.

Following landslips in 1891 and 1892, overhead girders were placed across the deepened lines to give additional support to the retaining walls.

Further reading

LELEUX R. *A regional history of the railways of Great Britain*, Vol.IX, *The east Midlands*. David and Charles, Newton Abbot, 1976, 51.

12. Grand Union Canal

HEW 1718

TQ 1877 & TQ 3680 to SP 0886 & SK 4930

In 1793 the Grand Junction Canal was opened from the River Thames at Brentford in London to a junction with the Oxford Canal at Braunston in Northamptonshire. The Engineers for the canal were James Barnes and William Jessop. During the construction of the 93 mile canal a number of major civil engineering works were constructed, including tunnels at Braunston (HEW 39), 2042 yd long and Blisworth (HEW 47) 3056 yd long, a large aqueduct over the River Nene at Cosgrove (HEW 44) and a 1 ½ mile long cutting at Tring, up to 30 ft deep.

By 1794, Leicester had been connected to the River Trent at Sawley, near Nottingham, through the canalization of the River Soar, by means of the Loughborough Navigation, opened in 1778 and the Leicester Navigation, opened in 1794. William Jessop was associated with both of these projects. In 1797, the Leicester and Northamptonshire Union Canal was opened between Leicester and Market Harborough, the original intention to connect Leicester with the Grand Junction Canal being frustrated by lack of finance.

At the end of 1799 two further canals opened, the Warwick and Birmingham and the Warwick and Napton canals, which together connected Birmingham to the Grand Junction via a short section of the Oxford Canal and thus completed a new route between Birmingham and London. The Warwick and Birmingham, engineered by William Felkin and Philip Henry Witton, included a short tunnel at Shrewley (HEW 707) of 433 yd and the great flight of 21 locks at Hatton (HEW 1037). William Felkin was also involved in the construction of the Warwick and Napton, with John Turpin and Charles Handley. Notable civil engineering works on this canal include the aqueduct over the River Avon at Warwick (HEW 641) and a long flight of 10 locks at Stockton. Later, an iron aqueduct was built to carry the canal over the Great Western Railway near Leamington Spa (HEW 495).

The gap remaining in the through route between London and Leicester was filled in August 1814 by the opening of the (old) Grand Union Canal which ran for 23 miles between the Grand Junction Canal at Norton Junction to the Leicester and Northamptonshire Union Canal at Foxton. James Barnes and Benjamin Bevan were the engineers for this canal which included the construction of tunnels at Husbands Bosworth of 1170 yd and Crick (HEW 1729), 1528 yds, and large flights of locks at Foxton (HEW 46) and Watford (HEW 42).

In 1894, the Grand Junction Company purchased the old Grand Union and the Leicester and Northamptonshire Union canals thus bringing the route from London to Leicester under one ownership. Later, in 1925, in the face of increasing competition from both rail and road transport, the Grand Junction initiated discussions with the Regents Canal, which ran from the Thames at Limehouse to the Grand Junction Paddington Branch, and the two Warwick canals with a view to some form of amalgamation. Following these negotiations agreement was reached in 1926 and under an Act of Parliament of 1928 the Grand Union Canal Company was formed and came into existence on 1 January 1929. The new company thus brought under one ownership a canal network of nearly 190 miles linking London with Birmingham and Leicester. The short section of the Oxford Canal between Braunston and Napton Junction forming part of the London to Birmingham route was not included.

In 1932 the Grand Union extended its network to the River Trent and beyond by purchasing the Leicester and Loughborough Navigations and the Erewash Canal which runs north from the River Trent to Langley Mill through the coalfields of the Erewash valley. Following this purchase the Grand Union Company controlled 227 miles of canal with 256 locks. Whereas the routes from London to Norton Junction and Braunston and from Foxton to the River Trent were suitable for broad boats the complete through operation of such craft was prevented by the existence of the narrow canals, between Norton Junction and Foxton and between Braunston and Birmingham.

Between 1932 and 1934 the Grand Union carried out extensive improvement works on the section between Braunston and Birmingham including the complete rebuilding of 51 locks to take 'broad' boats. This enabled broad craft to reach Birmingham from London but the facility was not greatly used. (For further details of the lock widening see Hatton Locks, section 8, HEW 1037.) Although the construction of the inclined plane lift at Foxton in 1900 (HEW 45) had permitted broad craft to proceed south from Foxton, the narrow locks at Watford, just north of Norton Junction, were not widened and the original narrow locks at Foxton were brought back into use in 1910 when the inclined plane was abandoned.

In 1993 to mark the 200th anniversary of the canal, the 140 mile towpath between Little Venice in London and Gas Street Basin at Birmingham was designated a National Waterways Walk, the first of its kind in the country.

Further reading

HADFIELD C. *The canals of the East Midlands.* David and Charles, Newton Abbot, 1981, 36–41, 79–92, 97– 135, 165–178, 187–188, 230–241, 239–247.

FAULKNER A. H. *The Grand Junction Canal.* David and Charles, Newton Abbot, 1972.

FAULKNER A. H. *The Warwick Canals.* Railway & Canal Historical Society, 1981, 5–18, 30–36.

13. Crick Tunnel

HEW 1729

SP 592 707 to SP 595 721

Crick Tunnel is the longer of two tunnels on the 20 mile summit pound of the old Grand Union Canal (HEW 1718). It is 1528 yd long and is brick lined throughout. The tunnel is 17 ft wide and 10 ft 6 in. high above the water

line with a truncated elliptical profile. There is no tow-path. The north and south portals are very similar with red brick wing walls which curve slightly with a stone coping. The brick arch is 13 in. thick.

The engineer for the tunnel was Benjamin Bevan. It was started on a different alignment in 1811 but trial borings revealed quicksands and other poor strata. A new tunnel was started on the present alignment in 1812. It was completed on the 29 July 1814 and opened for traffic on 9 August.

Falls of brickwork occurred in the tunnel in 1847, 1854 and 1867 following which repairs were carried out. In 1982 settlement was observed in one of the filled-in shafts. Major repair work carried out in 1986–87 included a new 10 m diameter shaft and 26 m of new tunnel of circular cross-section, 6.5 m (21 ft) in diameter, with pre-cast concrete lining.

Further reading

STEVENS P. A. *The Leicester Line.* David and Charles, Newton Abbot, 1972, 43, 55, 70–3, 74, 76, 77, 129, 149–52.

HADFIELD C. *The canals of the East Midlands.* David and Charles, Newton Abbot, 1981, 104–6.

14. Watford Locks

HEW 42

SP 593 688

Watford Locks are situated 2½ miles north of Norton Junction and lift the waterway through a vertical height of 54 ft to the 20 mile summit level of the canal.

There are seven locks. Lock number 1 (the bottom lock), 2 and 7 are normal single locks, but locks 3 to 6 form a staircase of four locks in which the top gates of one lock chamber form the bottom gates of the next higher lock. The staircase locks are provided with side ponds into which some of the water from the lock can be discharged and retained for re-use, thus economising on the use of water. The short lengths of canal between the single locks are widened on the west side to increase the amount of water held between the locks.

Both the locks at Watford and at the other end of the summit level at Foxton (HEW 46) were built to the narrow gauge and are only suitable for boats not exceeding 7 ft in width. This prevented the through passage of broad boats between the Grand Junction Canal and Leicester.

Watford Locks, staircase of four locks with side ponds on left, lock house above

Towards the end of the 19th century the Grand Junction Company, which had purchased the canal in 1894, proposed to build inclined planes at Watford and Foxton to bypass the narrow locks. However the proposed plane at Watford was never built and the locks were re-built as narrow locks in 1901–1902.

Further reading

FAULKNER A. H. *The Grand Junction canal*. David and Charles, Newton Abbot, 1972, 188, 222.

HADFIELD C. *The canals of the East Midlands*. David and Charles, Newton Abbot, 1981, 104–5.

15. Braunston Towpath Bridges

HEW 352

SP 532 660

During the 1829–34 improvements to the line of the Oxford Canal (HEW 1612) the junction between the Oxford Canal and the Grand Junction Canal at Braunston was relocated about ½ mile west of its original position. The new junction takes the form of a triangle of canals, the main line of the Oxford Canal forming the western side of the triangle and Braunston Branch connections to the Grand Junction Canal forming the north and east sides.

In order to carry the towpath over the junction two elegant cast-iron footbridges were constructed between the east and west sides of the junction and the central triangular island. These have spans of 50 ft, are 7 ft 9 in. wide and were cast by the Horseley Ironworks. They are made up from two semi-elliptical arch ribs incorporating the bridge parapet rails, each rib being cast in two halves with a joint at the crown of the arch. The deck is formed from cast-iron plates spanning between the arch ribs and covered with compacted earth. At the centre of the arch the bridges are 10 ft 8 in. above the water level. The two bridges are connected by a low semi-elliptical brick arch 40 ft long. The adoption of the semi-elliptical arch form gives increased headroom at the ends of the arch when compared with the equivalent segmental arch of the same span and rise.

The Braunston bridges are good examples of several bridges of similar design to be found on the Oxford Canal between Coventry and Braunston. They were mostly used to carry the towpath over the old loops of canal which remained when the line was straightened. Further examples can be found at Newbold-on-Avon (SP 482 778), Cathiron (SP 463 785) and Brinklow (SP 442 799). One further example, removed from the canal during construction works for the M6 Motorway at Sowe Common, has been restored and re-erected over the River Sherbourne at Spon End in Coventry (now at SP 327 788).

Further reading

HADFIELD C. *The canals of the East Midlands.* David and Charles, Newton Abbot, 1981, 159–162.

16. Braunston Tunnel

HEW 39

SP 557 655 to SP 575 652

Braunston Tunnel is 2042 yd long and carries the Grand Junction Canal under high ground to the east of the village of Braunston. Brick lined, it is 15 ft 10 in. wide and 12 ft 4 in. high.

The engineers for the tunnel were William Jessop and James Barnes, the Grand Junction Company's Engineer. The Contractors were Charles Jones and John Biggs, who agreed to build the tunnel for £9.10s. per yard.

During construction errors were made in setting out the line of the tunnel at the bottom of one or two of the working shafts and as a result of this it has a slight S-bend in the middle. Also, quicksands were encountered over a length of 328 yd of the tunnel which resulted in additional shafts having to be sunk.

Since the tunnel has no towpath boats were 'legged' through, but in 1870 a trial was made of an endless wire rope, driven by a steam engine, to haul boats through the tunnel. The trial was not a success and in 1871 a steam tug was introduced, the charge for towing boats through the tunnel varying from 1s. 6d. for a heavily loaded boat to 1s. for an empty craft.

HADFIELD C. *The canals of the East Midlands.* David and Charles, Newton Abbot, 1981, 110, 222, 280.

FAULKNER A. H. *The Grand Junction Canal.* David and Charles, Newton Abbot, 1972, 32–36, 178–179.

17. Holyhead Road (London to Coventry)

HEW 1214

TQ 314 831 to SP 365 757

This section, which is part of a continuous line of road between London and Holyhead runs for 85 miles between Islington, London and Toll Bar End at Coventry. A description of the remainder of the route can be found in *Civil engineering heritage Wales and western England* pp 9–11, 156–7, 175 .

The Royal Mail route to Ireland via Holyhead was, by the end of the 18th century, in a very bad state and was the subject of much criticism leading to the setting up of a Select Committee to investigate the road. In 1815 a Board of Commissioners was appointed with Thomas Telford as its engineer. He was assisted by William Provis and others in carrying out a survey between 1815 and 1817. The route via Coventry was selected for improvement mainly because it was already in use as the principal route to Manchester and Liverpool for part of the way.

The section between London and Coventry was left in the hands of the ten Turnpike Trusts which already administered it, with improvements being carried out to Telford's design and direction. The work proceeded over a number of years beginning in 1819 and progress is recorded in Telford's annual reports to the Commissioners. After problems encountered when handing back sections of road to the Turnpike Trustees immediately after improvement, the Commissioners obtained an Act which enables them to retain control over an improved section for two years before handing it back to the Trustees.

The original route of the road follows what is now the A1 from Islington to Finchley, the A1000 and A1081 to St Albans, the A5183 to the junction with the M1 and thence the A5 to Weedon Bec near Daventry. At Weedon the road turns onto the A45 through Dunchurch to Coventry. The form of construction depended on the original condition of the road but in general Telford's new road was 35 to 40 ft wide with a 30 ft carriageway and one or two footpaths about 5 ft wide. Extensive drainage of the foundations and surface of the road was provided, with side ditches, mitre and cross drains and simple gulleys in some places. A typical specification is quoted in Telford's *Life* which may be summarised:

> A level base with a bottom course of broken stones, 7 inches deep in the centre and 5 inches deep at the edge. An upper course of stone passing through a 2½ inch ring with a weight of not less than 6 ounces, to a depth of 6 inches. A gravel surfacing course 1½ inches deep.

Further reading

PENFOLD A. (ed.). *Thomas Telford: engineer*. Thomas Telford, London, 1980

Watling Street Pavement, following removal of vegetation

18. Watling Street Pavement

HEW 419

SP 657 563

In 1837 the Turnpike Commissioners carried out major improvements to the Holyhead Road between Towcester and Weedon in Northamptonshire in order to assist road coaches in competition with the London and Birmingham Railway (HEW 1092) then under construction. Major cuttings were made through the tops of two steep hills on this section of road and an embankment was built across the valley between. On the left side of each up-gradient a stone pavement was built. This pavement consisted of large granite blocks about 14 in. wide laid in two rows about 4 ft apart along which the wheels of the coaches ran. The space between the tracks was filled with granite setts.

It is recorded that the pavement was laid by Mr Edwards of nearby Heyford, Thomas Bishop of Fosters Booth and Thomas Reeve of Cold Higham. These men were under the direction of John Judkins, the Surveyor to the Commissioners.

The stone pavements were eventually covered by subsequent road works but during extensive improvements to the A5 road, carried out in 1954–57 a section of stone pavement was exposed. A short length of the pavement was preserved and re-laid in the east verge of the new road, just north of its junction with the B4525.

19. Charwelton Packhorse Bridge

HEW 393

SP 535 561

Travellers hurrying along the busy A361 Daventry to Banbury road through the village of Charwelton may notice on the south side of the road a narrow bridge which now carries the footpath; this is the oldest bridge in Northamptonshire. Originally intended to carry pack animals across a small stream forming part of the headwaters of the River Cherwell, the bridge has two spans of 7 ft and 6 ft 6 in. with pointed arches. The pathway is 4 ft wide flanked by low stone walls 12 in. wide and 15 in. high. With its approach ramps, the total length of the bridge is about 60 ft.

The bridge was built in the 13th century and has withstood the passage of time remarkably well although

Charwelton packhorse bridge

the upstream face has suffered in recent years from the spray thrown up by passing road traffic.

20. Blisworth Tunnel

HEW 47

SP 729 529 to SP 739 502

The longest canal tunnel in regular use on British Waterways, Blisworth Tunnel, is 3076 yd long and carries the Grand Union Canal under Blisworth Hill. The tunnel is lined with brickwork throughout, the thickness of the lining being 17 in. for the walls and roof and 13 in. for the invert. The waterway is 16 ft 6 in. wide at water level with a soffit height above water level of about 11 ft. There is no towpath.

During the construction of the tunnel 19 shafts were used, of which four were retained to act as ventilation shafts. After two men were asphixiated in the tunnel by the fumes from a steam boat in September 1861, further ventilation shafts were opened and there are now seven. The south portal of the tunnel is in its original condition and is built in red brickwork with a curved parapet wall. The north portal was rebuilt in blue brickwork in 1902–3. Both ends of the tunnel can easily be reached, the north (Blisworth) end by parking just above the tunnel entrance and walking down the access ramp and the south end by walking up the towpath from the Waterways Museum at

Blisworth Tunnel, south entrance

Stoke Bruerne. The mounds of earth deposited around the tops of the construction shafts can be clearly seen from the Blisworth to Stoke Bruerne road.

Construction of the tunnel took 12 years. A tunnel was begun in May 1793 on a different alignment but by 1796 the work had been halted by water problems. In June 1802 a new tunnel was started on the present alignment and was finally opened for traffic in March 1805. In order to allow traffic to bypass the tunnel during its construction, the Grand Junction Canal Company built first a toll-road and then a tramway over the top of Blisworth Hill. The tramway, design by Benjamin Outram, was about 3½ miles long, used horse-drawn wagons carrying about 2 tons and was completed in October 1800. The engineers for the tunnel were William Jessop and James Barnes.

The tunnel has always suffered from distortion of the lining because of poor ground conditions and it has recently been extensively repaired using a pre-cast concrete segmental lining, an example of which may be seen adjacent to the south portal.

Further reading

HADFIELD C. *The canals of the East Midlands.* David and Charles, Newton Abbot, 1981, 104, 110–111, 117, 221–222.

FAULKNER A. H. *The Grand Junction Canal.* David and Charles, Newton Abbot, 1972, 28, 42–57, 178–9, 204.

21. Barge weighing machine, Stoke Bruerne

HEW 603

SP 743 499

The tolls charged for the carriage of goods by canal were traditionally based on the ton-mile, that is the number of tons of goods carried by each boat multiplied by the distance carried in miles. This policy required the weight of goods carried by each boat to be measured at the toll office. Normally the measurement was made by gauging the boat, by measuring the distance between the waterline and the top of the boat side. Each boat would previously have been calibrated by measuring this distance for known tonnages and thus the tonnage could be calculated.

On a few canals, machines were installed to enable a complete boat with or without its cargo to be weighed.

Barge weighing machine, with weed cutting boat in cradle

One of these machines, which was built in 1836 for the Glamorganshire Canal, is preserved at the Waterways Museum at Stoke Bruerne in Northamptonshire. The machine is based on the principle of the double steelyard and uses a system of levers to enable the weight of the boat (many tons) to be balanced by relatively small weights placed on a scale pan. The machine has a cast-iron framework 32 ft long and about 12 ft wide and is placed in a dry dock which is initially filled with water. The boat to be weighed is floated into the dock and the water drained off, allowing the boat to settle onto a cradle about 50 ft long. The cradle is hung from inclined rods which bear on the ends of four levers, pivoted on the main framework of the machine. The four levers meet at the centre of the machine where they in turn bear on the end of a fifth lever which is set at right angles. This lever has a scale pan hanging from the end into which the balancing weights are placed. One pound of weight placed in the scale pan will balance one hundredweight on the cradle.

Further reading

PAGET-TOMLINSON E. W. *Canal and river navigations*. Waine Research Publications, 1978, 26.

22. Cosgrove Aqueduct

HEW 44

SP 801 418

One of the major obstacles facing the builders of the Grand Junction Canal (HEW 1718) was the crossing of the Great Ouse River near Wolverton. In December 1799 James Barnes submitted a scheme for an aqueduct across the river but because of the lengthy delay in construction of the aqueduct a temporary level crossing of the river was brought into use in September 1800. The canal descended to the level of the river by means of four locks on each side, traces of which may still be seen on the west side of the present aqueduct.

In June 1802, William Jessop submitted a scheme for a stone arched aqueduct and in December the tender of Thomas Harrison and others was accepted. There were several difficulties during construction but the structure, with three semicircular arches, was opened on Monday 26 August 1805.

In January 1806 a section of the canal embankment

Cosgrove Aqueduct

subsided and on 18 February 1808 the aqueduct itself collapsed. The old locks were brought back into use until a temporary timber trough was opened in June.

The present cast-iron trough aqueduct was designed by Benjamin Bevan and was cast by Reynolds and Company of Ketley Ironworks. Work began in September 1809 and the aqueduct was opened on 22 January 1811. It has two spans, and the trough, 15 ft wide and 6 ft 6 in. deep, has a total length of 101 ft. The towpath is on the east side of the trough, level with the top, and is supported by inclined struts. The trough is made up from a base-plate and two side plates with bolted joints. There are nine individual sections in each span, the side joints being set at varying angles. The top of the trough is about 35 ft 6 in. above the level of the river. The central pier is of masonry with a decorative stone coping. The abutments are faced with blue brickwork which was added during repair work in 1921.

Further reading

FAULKNER A. H. *The Grand Junction Canal.* David and Charles, Newton Abbot, 1972, 60–70, 204.

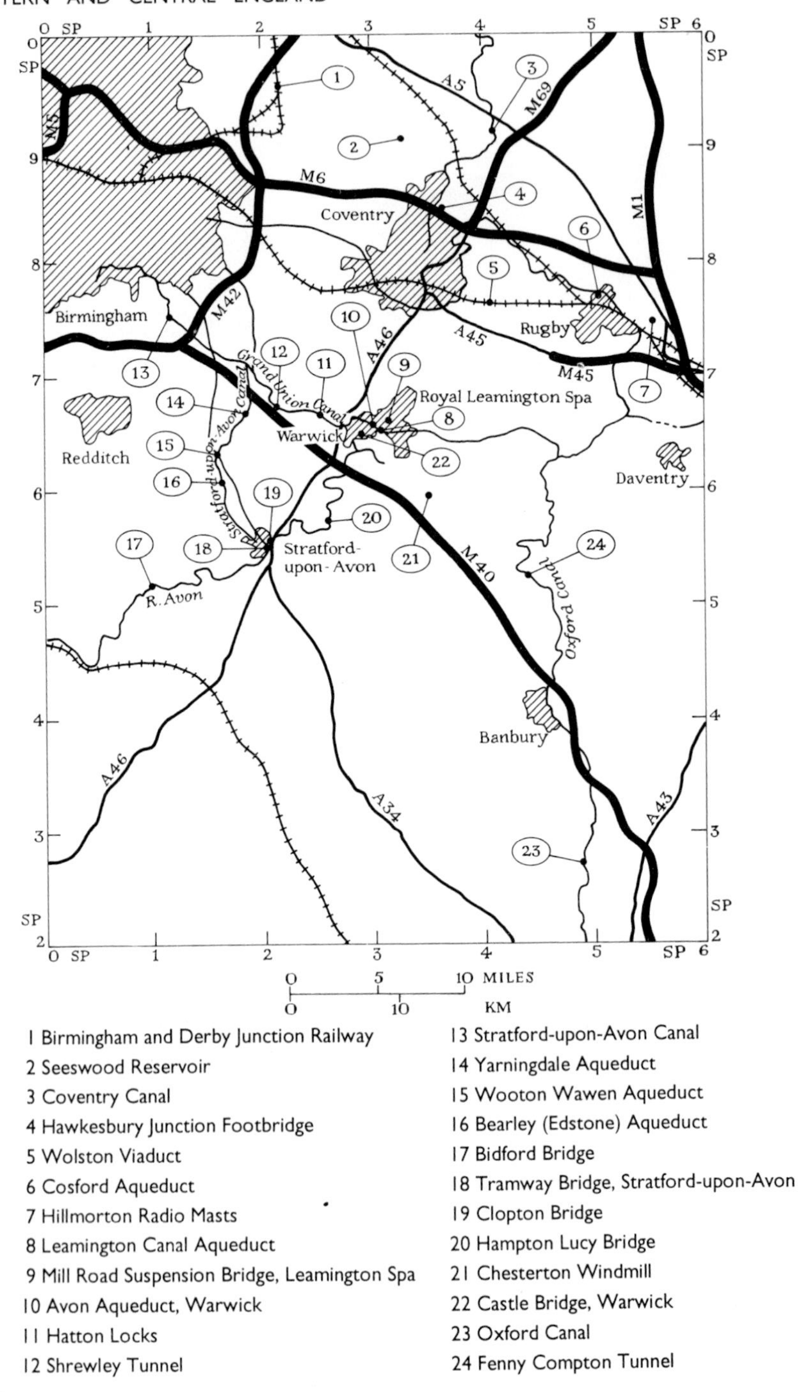

1 Birmingham and Derby Junction Railway
2 Seeswood Reservoir
3 Coventry Canal
4 Hawkesbury Junction Footbridge
5 Wolston Viaduct
6 Cosford Aqueduct
7 Hillmorton Radio Masts
8 Leamington Canal Aqueduct
9 Mill Road Suspension Bridge, Leamington Spa
10 Avon Aqueduct, Warwick
11 Hatton Locks
12 Shrewley Tunnel
13 Stratford-upon-Avon Canal
14 Yarningdale Aqueduct
15 Wooton Wawen Aqueduct
16 Bearley (Edstone) Aqueduct
17 Bidford Bridge
18 Tramway Bridge, Stratford-upon-Avon
19 Clopton Bridge
20 Hampton Lucy Bridge
21 Chesterton Windmill
22 Castle Bridge, Warwick
23 Oxford Canal
24 Fenny Compton Tunnel

8. Warwickshire and North Oxfordshire

In 1974 the County of Warwickshire lost much of its industrially developed land to the newly created County of the West Midlands which included Birmingham and Coventry, hitherto parts of Warwickshire. Since that time the County has been of a predominantly rural nature, with Warwick, Leamington Spa, Stratford-upon-Avon, Rugby and Nuneaton as major towns. North Oxfordshire is of a similar nature, with Banbury the only large town in the area.

Geographically, the area is one of contrast with the centre of the area lying in the valley of the River Avon, which rises in the north-east of the Northamptonshire border. In the north-west, the land rises to the Birmingham plateau whereas to the east and south-east lies a ridge of high ground which marks the northernmost end of the Cotswold Hills, running roughly along the border with neighbouring Northamptonshire. In the south, the River Cherwell rises in these hills and flows down through Banbury towards Oxford and the Thames.

The area lies squarely on the line joining Birmingham to London and has been crossed by most of the major transport routes connecting two of Britain's largest cities. The Roman Watling Street skirts the area to the east and north but later routes crossed the area more directly, including the Oxford and Coventry Canals, followed slightly later by the Warwick and Birmingham and Napton Canals, both eventually to form part of the Grand Union network. The route between Rugby, Coventry and Birmingham was chosen for one of Britain's earliest major railways, the London and Birmingham, and more recent developments have seen the construction of the M6 and M40 Motorways across the area.

Because of this predominance of transport routes, the majority of the sites described in this section will be found to relate to roads, railways and canals and their associated structures.

1. Birmingham and Derby Junction Railway

HEW 1667

SK 362 356 to SP 086 873

Across the fields between Hampton-in-Arden and Whitacre to the east of Birmingham the track of an abandoned railway can still be traced. Although the line may appear to have been one of the multitude of minor country railways with which Britain was so liberally provided, it was in fact for the three years between 1839 and 1842 a major link in the national rail network.

In 1836 the Birmingham and Derby Junction Railway Company (B & DJ) obtained an Act of Parliament empowering the company to construct a railway from Derby via Burton-on-Trent and Tamworth to a junction with the London and Birmingham Railway (L & B) at Stechford and thence to run via the L. & B.'s tracks to Curzon Street Station in Birmingham. However, the B & DJ soon changed its plans and instead of constructing the line from Whitacre to Stechford, built the 6 mile line from Whitacre to Hampton-in-Arden where it joined the London and Birmingham with a junction facing London. Thus on 12 August 1839 a new, albeit rather roundabout, route was opened up from Derby and the north to London.

Between Derby and Whitacre the route of the railway, surveyed by George Stephenson, followed the valleys of the Rivers Trent and Tame and few major engineering works were required, apart from the viaduct over the river at Wychnor and a 19 arch viaduct at Tamworth (HEW 886), described in a contemporary reference as '... scarcely surpassed for size and beauty...'. John Birkinshaw was placed in charge of construction.

Unfortunately relations between the B & DJ and the L & B were not good and the B & DJ was soon seeking to construct an independent line into Birmingham. Abandoning its original intention to join the L & B at Stechford, the Company built a new line from Whitacre along the valley of the River Tame to a new terminus at Lawley Street, near to Curzon Street. This line was opened on 10 February 1842, the total distance between Derby and Birmingham being 41½ miles. In April a spur was opened between the B & DJ and the Grand Junction Railway near Landor Street in Birmingham and traffic between

Derby and Curzon Street via Hampton ceased. Also by this time, traffic from Derby to London was mainly using the Midland Counties Railway between Derby and the L & B at Rugby, avoiding the roundabout route through Hampton. Although the Whitacre to Hampton line remained in existence for many years, its importance was greatly diminished and in 1917 it was singled and closed to passenger traffic. It was finally closed to all traffic in 1930.

The main line of the B & DJ remains in full use providing an important link in the route between the south-west and the north-east. The company amalgamated with the Midland Counties Railway and the North Midland Railway in 1844 to form the Midland Railway.

Further reading

WHISHAW F. *The railways of Great Britain & Ireland*, 1840, reissued David and Charles, Newton Abbot, 1969, 15–18.

LELEUX R. *A regional history of the railways of Great Britain, Vol. IX, The east Midlands*. David and Charles, Newton Abbot 1976, 169–70, 175–77.

2. Seeswood Reservoir

HEW 1730

SP 329 905

Travellers using the B4102 road from Nuneaton to Meriden in Warwickshire will notice a short section of straight road running alongside a lake about 2 miles south of Nuneaton. In fact the road runs along the top of the impounding dam of Seeswood Reservoir, claimed to be the oldest canal reservoir to be built in Britain.

The reservoir was built in 1764 to supply water to the extensive network of private canals built on his Arbury estate by Sir Roger Newdigate, the fifth Baronet. The dam is about 460 ft long and 33 ft wide at the top with a maximum height of about 22 ft. A feeder channel just over 1 mile long ran from the reservoir to the canal system and in 1777 the feeder was made navigable and a lock was constructed through the south-west end of the dam to enable boats to enter the reservoir itself. Traces of the lock can still be seen but at some time the north-east wall of the lock chamber was demolished and the resulting slope concreted to provide a spillway.

Following Sir Roger's death in 1806 the Arbury canal

system rapidly fell into disuse but Seeswood Reservoir has survived and is now extensively used for fishing.

Further reading

WEAVER P. The Arbury canals. *Journal of the railway and Canal Historical Society*. 1970, Jan. 1–7, Apr. 28–33.

3. Coventry Canal

HEW 1090

SK 140 140 to SP 333 796

The Coventry Canal was one of the earliest canals to be built in Britain and forms an important part of the link between the River Trent and the River Thames. A narrow canal built to take boats 72 ft long and 7 ft beam, it runs 38 miles south from its junction with the Trent and Mersey Canal (HEW 1135, SK 140 140) at Fradley to a terminus at Bishop Street Basin (HEW 1527, SP 333 795) near the centre of Coventry. At Hawkesbury Junction, north of Coventry, there is a junction with the Oxford Canal which completes the link to the Thames.

The canal was designed by James Brindley and construction began in 1768 starting at the Coventry end and working northwards. By the time Atherstone, 16½ miles from Coventry, had been reached in 1771 the authorised capital of £50 000 had been used up and construction ceased. For several years the canal existed as an isolated length until in 1777 it was joined by the Oxford Canal at Longford, north of Coventry. The remaining section of the Coventry's line, from Atherstone to Fradley, was finally completed by the joint efforts of no less than three canal companies. The section from Atherstone to Fazeley was built by the Coventry company, from Fazeley to Whittington Brook by the Birmingham & Fazeley Canal company (later amalgamated with the Birmingham Canal Company) and the final section from Whittington Brook to Fradley by the Trent and Mersey company. The whole canal was finally opened to through traffic on 13 July 1790, 22 years after its commencement.

The section from Whittington Brook to Fradley was subsequently purchased by the Coventry company but the intervening section from Fazeley to Whittington Brook remained in the ownership of the Birmingham company, thus splitting the Coventry's line into two parts.

The route of the canal, from north to south, follows the valley of the River Tame as far as Tamworth where it crosses the river by an aqueduct. It then climbs up the valley of the River Anker by thirteen locks (including a flight of eleven at Atherstone) and then runs level for 16 ½ miles to Coventry. The junction with the Oxford Canal has an interesting history. Due to a complicated dispute over tolls the junction was originally built at Longford, although this meant that the two canals ran parallel for a considerable distance before joining. In 1785 the junction was relocated at Hawkesbury which removed most of the parallel canals. In 1804 a junction between the Coventry Canal and the Ashby-de-la-Zouch Canal at Marston, near Bedworth, was opened.

HADFIELD C. *The canals of the East Midlands.* David and Charles, Newton Abbot, 1981, 15–17, 19, 22–24, 142–156.

SIVEWRIGHT W. J. *Civil Engineering Heritage Wales and western England.* Thomas Telford, London, 1986,159–160.

Below: Hawkesbury Junction Footbridge

4. Hawkesbury Junction Footbridge

HEW 1111

SP 363 846

When the junction between the Coventry Canal (HEW 1090) and the Oxford Canal (HEW 1612) was relocated at Hawkesbury, north of Coventry, in 1837, an elegant cast-iron footbridge was cast by Handysides of Derby at the

Britannia Foundry to the design of Mr J. Sinclair to carry the towpath of the Coventry Canal across the junction.

The bridge has a span of 60 ft with a semi-elliptical arch of 7 ft 5 in. overall width. The arch is formed of two main ribs each made up from three sections with outer cover plates bolted through both segments. The main rib is an L-section 1 ft 3 in. deep and 2 in. thick with an integral parapet of X-lattice form 4 ft high. The bridge deck is formed from curved cast-iron plates bolted to the rib flanges and paved with red brick blocks with raised courses.

5. Wolston Viaduct

HEW 767

SP 409 761

A low viaduct, carrying the line of the London and Birmingham Railway (HEW 1092) across the River Avon and the Brandon to Wolston road, was built in 1837 for the opening of the railway in 1838.

Designed by Robert Stephenson, the viaduct has nine semi-elliptical arches of 24 ft span with three smaller semicircular arches of 10 ft span at each end. The arches are built in brickwork with stone facings, as are the supporting piers. At the west end, two of the smaller arches have been bricked up and the embankment extended, partially concealing them. At the foot of the parapet wall is a decorative ogee stone moulding.

Although the development of nearby land and the growth of vegetation have obscured the fine view of the viaduct, illustrated in the engraving by Bourne, the structure still stands virtually unchanged after 150 years of use.

The viaduct is very similar in design and construction to the much larger structure at Wolverton (HEW 152, section 6).

Further reading

ROSCOE T. *The London and Birmingham Railway*. Charles Tilt, 1838, 116–7.

6. Cosford Aqueduct

HEW 1720

SP 503 771

Between 1829 and 1834 extensive improvements were made to the line of the Oxford Canal between its junction with the Coventry Canal at Hawkesbury Junction and Braunston. North-west of Rugby the original line of the canal made a lengthy detour to the north up the valley of the River Swift which it crossed just south of the village of Cosford by a small brick aqueduct. During the improvements a new section of canal was constructed to avoid this detour and in order to carry the new line of the canal over what was then the road from Rugby to Lutterworth a cast-iron trough aqueduct was constructed.

The canal is carried in a cast-iron trough 15 ft wide and 6 ft deep. The trough spans 23 ft 6 in. between the brick abutments and is made up of base and side castings with flanged and bolted joints. The side joints of the trough are inclined to the vertical, the inclination increasing with increased distance from the centre of the span. The base of the trough was originally supported by four cast-iron segmental arch ribs but three of these ribs have been fractured in collisions with vehicles exceeding the 13 ft 3 in. headroom under the trough and have been replaced by modern steel beams. On both sides of the trough is an area 8 ft 9 in. wide supported by iron plates spanning between the top of the trough sides and a pair of elegant cast-iron segmental arches with a span of 23 ft 6 in. and a rise of 6 ft 6 in. The spandrels of the arches have five radial struts. The towpath of the canal is on the south side of the aqueduct. The brick wing walls curve through 90° before sloping down to ground level and there is a cast-iron parapet rail on both sides of the aqueduct.

The structure was designed by William Cubitt, Consultant to the Oxford Canal Company and Frederick Wood, the Company's Engineer. The new line of the canal was fully opened to traffic on 13 February 1834.

Further reading

HADFIELD C. *The canals of the East Midlands*. David and Charles, Newton Abbot, 1981, 160–161.

7. Hillmorton Radio Masts

HEW 763

SP 553 745

In 1924 the Wireless Telegraphy Commission and the Post Office Engineering Department sited the 'worldwide' radio transmitting station on a 900 acre site, 3 miles east of Rugby. The general ground level of the site is 340 ft above sea-level.

To support the main transmitting aerials, twelve steel cable-stayed masts were constructed. The masts are arranged in a full ellipse of eight masts and a half ellipse of six masts, two masts being common to both loops. The masts are spaced horizontally about 1300 ft apart and the long axis of the system is aligned in the north-west to south-east direction.

The masts, which were built by the Head Wrightson Company, are 820 ft high and are of steel lattice construction. The 'K' lattice is triangular in plan with sides 10 ft long. Each mast is supported on a cast steel shoe, which is itself mounted on a granite insulator block of 5 ft 6 in.[3] The total dead weight of each mast and its stays is 200 tons but allowing for the tension in the supporting stays the total downward force on the mast base is 400 tons. Access to the top of the masts is gained by means of a lift running inside the mast structure.

Further reading

ANGWIN A. S. and WALMESLEY T. Rugby radio station. *Proc. Instn Civ. Engrs* 1925–1926, 22, Part 1, 201.

8. Leamington Canal Aqueduct

HEW 495

SP 303 653

The section of the Warwick and Napton Canal between Warwick and Leamington Spa was opened in 1800. In 1851 construction started of a railway between the two towns, part of the Birmingham and Oxford Junction railway, which was leased by the Great Western Railway in 1848. The railway passes underneath the canal and to carry the waterway a cast-iron aqueduct was built by Messrs. Peto and Betts.

The canal is carried in a cast-iron trough 18 ft 9 in. wide which crosses the railway on a skew angle of 45° with four equal spans of 21 ft on the skew. The trough is supported over each span by six segmental cast-iron arch ribs. The form of the structure is complex, the arch ribs being cast in the form of double cantilevers which engage with lugs cast on the bottom side of the trough base so that the trough acts as the centre of the arch. On the outer pair of ribs the arch is completed by extra arch sections bolted alongside the trough.

The outer sections of the structure which extend beyond the trough for about 7 ft on both sides are supported

Leamington Canal Aqueduct

by brick arches which spring from the same 2 ft 10 in. wide brick piers as the cast-iron arches.

Further reading

McDermot E. T. and Clinker C. R. *History of the Great Western Railway. Vol. 1.* Ian Allan, London, 1982, 166–7.

9. Mill Road Suspension Bridge, Leamington Spa

HEW 1585

SP 321 656

Mill Road Bridge, which carries a footpath over the River Leam in Jephson Gardens, Leamington Spa, was opened in June 1903 and is a steel suspension bridge with a span of 100 ft. The deck is supported by suspension rods which run from the top of steel lattice towers 19 ft high to the third points of the span. Two rods run from the tower top to the nearer third point and a single rod to the further third point. Thus each third point is supported by three rods, two from the nearer tower and one from the further tower. The long rods are 2 in. in diameter and the shorter rods are 1½ in. diameter. On the landward side all three rods run from the tower top to an anchorage. The suspension rods are attached to vertical hangers level with the top of the parapet, the hangers running down to the main deck beams.

Leamington Suspension Bridge

The deck is formed of three Warren lattice girders on each side of the bridge with cross beams. The footway is formed of steel plates covered with a thin layer of asphalt.

Underneath the bridge the river falls over a curved, stepped weir with the gardens forming a pleasing backgound.

Further reading

STORRIE J. *William Louis De Normanville Engineer, Architect and Inventor.* Weir Books, 48–51

10. Avon Aqueduct, Warwick

HEW 641

SP 301 655

The most westerly section of the Grand Union Canal from Birmingham to Napton Junction on the Oxford Canal was actually built as two separate canals, the Warwick and Birmingham Canal, opened in 1799 and the Warwick and Napton Canal, opened to traffic in March 1800.

Only a short distance from its junction with the Warwick and Birmingham, the Warwick and Napton encountered one of its major obstacles, the crossing of the River Avon. Approaching the river on a high embankment, the canal crosses the river by a three-arch stone aqueduct, the segmental arches having a span of 43 ft and a rise of 14 ft. The two piers are 9 ft wide with substantial cutwaters. Over the aqueduct the canal waterway is 16 ft 3 in. wide with a water depth of 5 ft 2 in. Together with its curved wing walls, the overall length of the structure is about 230 ft and it carries the waterway about 30 ft above the river level. It was designed by Charles Handley assisted by Henry Couchman.

On the towpath side of the aqueduct is a concrete parapet wall 4 ft 3 in. high, dated 1909, which has been cast directly on the original stone coping.

Further reading

HADFIELD C. *The canals of the East Midlands*. David and Charles, Newton Abbot, 1981, 173.

11. Hatton Locks

HEW 1037

SP 264 655 to SP 241 669

In order to carry the line of their canal down into the valley of the River Avon, the Warwick and Birmingham Canal Company built an impressive flight of 21 locks at Hatton, east of Warwick. The locks are spread over a length of canal of 3400 yd from Hatton Top lock (No. 46) to Hatton Bottom lock (No.26). The locks are closely spaced at the top of the flight, the spacing increasing towards the bottom of the flight as the land flattens out into the river valley.

Hatton Locks

When they were opened in December 1799 the locks were built to the narrow gauge for boats 72 ft long and 7 ft wide. The total fall of the lock flight is 146 ft 6 in., giving an average fall for each lock of 7 ft. The Engineers for the locks were William Felkin and Philip Henry Witton.

In 1929 the Warwick and Birmingham Canal became part of the Grand Union Canal Company and in 1932–34 the locks at Hatton were completely rebuilt to take 'broad' boats 83 ft 6 in. long and 15 ft wide. In the majority of cases the old narrow lock chamber was retained alongside the new lock to act as an additional reservoir (for the short pounds) and as an overflow weir. Where the old lock chamber is on the towpath side of the new lock it has been covered by a concrete slab.

There is a similar flight of nine locks at Stockton on the other side of the Avon valley.

Further reading

HADFIELD C. *The canals of the East Midlands.* David and Charles, Newton Abot, 1981. 165–6, 241.

FAULKNER A. *The Warwick canals.* Railway and Canal History Society, 1985, 1985, 6,7,22,46–8,57,65–6.

12. Shrewley Tunnel

HEW 707

SP 214 672

Shrewley Tunnel, 433 yd long, was opened in 1799 to carry the Warwick and Birmingham Canal (later part of the Grand Union) under a ridge of high ground at the village of Shrewley in Warwickshire.

The tunnel is brick lined, 16 ft 2 in. wide and 12 ft 9 in. high. The exceptional feature of the work, which is otherwise typical of many canal tunnels of the period, is that at its west end, the towpath, having left the line of the canal to pass over the top of the tunnel, then passes through a short tunnel of its own. The towpath tunnel is about 176 ft long, 9 ft wide and 8 ft high. It has vertical walls and a semicircular arched roof and rises on a steepening gradient. The floor of the tunnel is paved with brick with raised courses which afforded a grip to the hooves of horses. Where the towpath enters its tunnel it is 15 ft above the level of the canal, being supported by a retaining wall.

On leaving the towpath tunnel, the path crosses the

main village street then follows the line of the canal tunnel across the fields to rejoin it at its eastern portal.

Shrewley Tunnel

13. Stratford-upon-Avon Canal

HEW 1653

SP 054 794 to SP 205 548

In March 1793 an Act was passed for the construction of a canal from a junction with the Worcester and Birmingham Canal at Kings Norton in Birmingham to the town of Stratford-upon-Avon. The act did not include provision for any connection with the River Avon at Stratford. William Clowes, the Engineer, began cutting the canal from Kings Norton, the line running east, then south-east, until it reached Hockley Heath in May 1796. Here cutting ceased for lack of money but was restarted in 1800 under Samuel Porter as far as Kingswood where cutting again ceased. Restarted again in 1812, the canal proceeded south from Kingswood to reach Stratford in June 1816. Here a connection with the River Avon, authorised by an Act of 1815, was made. At Kingswood there was also a short branch canal connecting with the line of the Warwick and Birmingham Canal.

The canal was built for narrow boats and is very heavily locked, there being 55 locks in the 25 ½ miles from Kings Norton to Stratford, most of them concentrated in a

Stratford-upon-Avon Canal, Kingswood Junction

length between Hockley Heath and Stratford. Earlswood Lakes (SP 113 742), which provide a feeder reservoir for the canal, were built in 1821–22. The three seperate pools, Windmill, Engine and Terry's are retained by earth embankments and have a total capactiy of 210 million gallons.

There are several notable features on the Stratford Canal, including the use of 'split' cast-iron footbridges at the tail of the locks to enable the tow rope to be dropped through the bridge. There are three notable cast-iron aqueducts south of Kingswood, at Bearley (HEW 281), Wootton Wawen (HEW 282) and Yarningale (HEW 655).

In 1960 the southern section of the canal between Kingswood and Stratford, then derelict, was taken over by the National Trust and restored to full use, one of the first examples of the restoration of a British canal. In 1988 the ownership of this section was returned to British Waterways.

Further reading

HADFIELD C. *The canals of the East Midlands.* David and Charles, Newton Abbot, 1981, 179–181, 208.

HADFIELD C. and NORRIS J. *Waterways to Stratford.* David and Charles, Newton Abbot, 1968.

14. Yarningale Aqueduct

HEW 655

SP 184 664

During the construction of the southern section of the Stratford-upon-Avon Canal between Kingswood and Stratford two cast-iron trough aqueducts were constructed at Bearley (HEW 281) and Wootton Wawen (HEW 282). About 18 years after the opening of this section of canal, a burst on the nearby Warwick and Birmingham Canal released a torrent of water into a stream which swept away the aqueduct at Yarningale Common on the Stratford Canal. A third cast-iron trough aqueduct was constructed at Yarningale to replace the damaged aqueduct, using the same design as that previously used at Bearley and Wootton Wawen.

The smallest of the three aqueducts, it is 42 ft long and crosses the stream in a single span between brick abutments and leads directly into a lock at its southern end. The towpath is level with the base of the trough and is carried on an extension of the trough base plate. An inscription records that it was cast by the Horseley Ironworks in 1834.

Further reading

HADFIELD C. and NORRIS J. *Waterways to Stratford*. David and Charles, Newton Abbot, 1968, 101–103.

15. Wootton Wawen Aqueduct

HEW 282

SP 159 630

A cast-iron trough aqueduct carries the Stratford-upon-Avon Canal over the A34 Stratford–Birmingham road at Wootton Wawen. Three brick piers divide the aqueduct into four unequal spans, the total length of the structure being about 100 ft. The trough is made up of cast-iron base and side plates with flanged bolted joints but unlike the similar structures at Bearley (HEW 281) and Yarningale (HEW 655) the lengths of the trough side sections are not equal, varying from 7 ft to 9 ft in length. The trough is given additional support by two cast-iron beams and the towpath is carried level with the base of the trough on the west side on an extension of the trough base plate.

The aqueduct carries an inscription:

> 'This aqueduct was erected by the Stratford Canal Company in October 1813. Bernard Dewes Esq., Chairman, W. James Esq. Depy. Chairman, W.Whitmore Engineer.'

Further reading

HADFIELD C. *The canals of the East Midlands*. David and Charles, Newton Abbot, 1981, 179–181.

16. Bearley (Edstone) Aqueduct

HEW 281

SP 162 609

This aqueduct is the largest of three similar aqueducts built on the southern section of the Straford-upon-Avon Canal and is the second longest cast-iron canal aqueduct in Britain.

Designed by William Whitmore and opened to traffic in 1816, the aqueduct has a cast-iron trough supported by cast-iron beams spanning between slender brick piers. With 14 spans of 34 ft 3 in. the total length of the structure is 479 ft which is exceeded only by Thomas Telford's 1027 ft long Pont Cysyllte aqueduct. (HEW 112)

The trough is 8 ft 10 in. wide and 5 ft deep, the towpath being carried alongside and level with the bottom of the trough on an extension of the trough base plate to give an overall width of 13 ft. In section the aqueduct is made up from four flat plates (one base plate, two trough side plates and one towpath side plate) all with flanged edges and bolted joints. The trough side plates are 14 ft long and the base plates are half this length, the joints in the side and base plates being staggered. Cast-iron beams give additional support to each span.

Top: Wooton Wawen Aqueduct

With a maximum height of 33 ft the aqueduct crosses a road and a stream and also the line of the ex-Great

Western Stratford to Birmingham railway, opened in 1906, the twin tracks diverging to pass either side of one of the piers.

In placing the towpath at a low level, alongside the trough, Whitmore was following the example set by Thomas Telford in his design for the first major cast-iron aqueduct at Longdon-upon-Tern in Shropshire, (HEW 280) built in 1796. In Telford's later aqueduct, Pont Cysyllte, he carried the towpath at a higher level, cantilevered over the water in the trough thus enabling a wider trough to be provided and lessening the 'piston' effect of the small clearance between the sides of the boat and the trough sides. It would seem that either Whitmore was unaware of this improvement in design or chose to ignore it.

17. Bidford Bridge

HEW 1021

SP 099 517

This is a late medieval stone bridge over the River Avon carrying the Bidford-on-Avon to Badsey B4085 road. It has obviously been heavily altered down the years but is believed to date from the 15th or 16th century. There are eight arches of varying styles, arches 1 and 2 (from the north bank) being irregular and pointed, arches 3, 5, 6 and 7 being segmental, arch 4 being semicircular and the eighth and southernmost arch is pointed. The fourth arch is significantly higher than the others. The spans of the arches are similarly variable, from 11 ft 8 in. to 17 ft. The present navigation arch is number 6.

The roadway over the bridge varies in width with a maximum of 12 ft 8 in.; the total length of the bridge is 203 ft over the river and about 320 ft overall including the approach ramps. There are seven triangular cutwaters on the upstream face of the bridge which also incorporate pedestrian refuges.

It is recorded that in 1537 one Robert Cannynge left the sum of five shillings for the repair of 'ye brigge of bydeford'.

Further reading

JERVOISE E. *The ancient bridges of Wales and western England*. Architectural Press, London, 1933, 165–6.

18. Tramway Bridge, Stratford-upon-Avon

HEW 868

SP 205 548

In 1821 an Act of Parliament was passed which authorised the construction of a tramway linking Stratford-upon-Avon and Moreton-in-Marsh. This line was the only part actually built of the much greater scheme for a Central Junction Railway from Stratford to London, proposed by William James.

The line started at the canal wharf in Stratford and in order to cross the River Avon a brick bridge was built, parallel to and a little further downstream from the ancient Clopton Bridge (HEW 679). The bridge has nine elegant semi-elliptical arches of 30 ft span, the last two arches at the south end of the bridge being over dry land. The width of the bridge is 10 ft 8 in. between the brick parapet walls and the total length of the structure is about 350 ft. The bridge is now used as a pedestrian way.

The 17-mile long route of the Stratford and Moreton Railway, opened in 1826, was leased in 1844 to the Oxford, Worcester and Wolverhampton Railway and remained in service until the end of the 19th century. Horse traction was used throughout, steam locomotives being prohibited from the line where it ran alongside public roads, which it did for much of its length. The gauge was 4 ft 8½in. The railway and bridge were built by J. U. Rastrick, although William James was closely concerned with the survey and design of the line. Thomas Telford and George Stephenson were also consulted by the Tramway Company during the planning and construction stages.

Further reading

NORRIS J. *The Stratford and Moreton Railway*. Railway and canal historical Society, 1987.

19. Clopton Bridge

HEW 679

SP 206 548

Built by Sir Hugh Clopton in about 1480–90 to replace an earlier bridge, Clopton Bridge carries the main A46 road over the River Avon in Stratford-upon-Avon. The bridge has 14 pointed arches varying in span from 18 ft 6 in. to 19 ft. Overall, the bridge is 482 ft long.

Clopton Bridge, Stratford

In 1814 the bridge was widened by adding new arches on the upstream (north) side. The piers of the new arches are extensions of the existing piers but the spans of the new arches are larger than the spans of the original arches, varying from 25 ft 5in. to 29 ft 2 in. The 21 ft width of the bridge was still considered inadequate for in 1827 a footway was added to the bridge, again on its upstream side. The footway, which is 3 ft 9 in. wide was constructed from cast-iron plates which are supported on cast-iron brackets fastened to the face of the bridge and supported on the cutwaters of the piers. There is an octagonal toll house at the town end of the bridge.

An inscription on the wall of the bridge records that:

> Sir Hugh Clopton Knight Lord Mayor of London built this bridge in the reign of King Henry Seventh.

Despite its later additions, the bridge, especially when seen from the south, presents an appearance largely unchanged for 500 years.

Further reading

Victoria County history of Warwickshire, **III**, OUP, 1945, 224.

JERVOISE E. *The ancient bridges of Wales and western England*. Architectural Press, 1933, 163–4.

20. Hampton Lucy Bridge

HEW 1050

SP 258 572

In 1829 a new cast-iron bridge, fabricated at the Horseley Ironworks, was built over the River Avon at Hampton Lucy. It was paid for by the Reverend John Lucy, Rector of Hampton Lucy, and replaced a ford and raised wooden causeway for foot passengers. The Lucy family were the owners of nearby Charlecote House.

The bridge has a cast-iron segmental arch of 60 ft span and 7 ft rise with four ribs, each rib being made up from four segments. The arch ribs, which are 1 ft 8 in. deep and 4¼ in. wide have conventional flanges on one side but are flat on the other side and are placed so that the flat side is towards the centre of the bridge. The arch is braced laterally by flat horizontal bars at the segment joints and by circular rods. The spandrel is filled with an X-lattice of T-shaped bars with a decorative moulding.

The abutments are unusual, being of well-worn sandstone ashlar masonry, pierced with five narrow pointed arches which are 2 ft 2 in. wide and vary in height from 5 ft to 5 ft 6 in., beyond which are curved wing walls.

The bridge has recently been refurbished and now carries single file traffic, with a weight limit of 7½ tonnes

Further reading

Victoria County history of Warwickshire. **3**, OUP, 1945, 101.

21. Chesterton Windmill

HEW 867

SP 348 594

Chesterton Windmill stands on high ground to the east of the Foss Way, overlooking the earthworks of the nearby Roman station. The mill is a tower mill of most unusual form, built in 1632 for Sir Edward Peyto, and its design has been attributed to Inigo Jones. Architecturally it is of considerable merit and the machinery is of outstanding interest.

The mill, circular in plan with a diameter of 22 ft 9 in. and a height of 36 ft, is built of local hard limestone, carefully coursed and generally of ashlar form. The upper part of the tower, containing the machinery of the mill, is supported on six semicircular moulded arches on approximately rectangular piers, the outer faces of which are arcs of circles radiating from a common centre. A sandstone string course surmounts the arches and runs round the tower below the four windows. The rotating cap of the mill is in the form of a shallow dome upon which the sails are mounted.

The machinery of the mill follows the normal pattern

Chesterton Windmill

with the inclined wind shaft of the sails driving a vertical main shaft which in turn drives two millstones from the spur wheel.

Further reading

BROWN R. J. *Windmills of England*. Robert Hale, London, 1976, 224–5.

22. Castle Bridge, Warwick

HEW 1089

SP 288 647

In 1788 the Earl of Warwick obtained an Act of Parliament to replace the Great Bridge over the River Avon below the walls of Warwick Castle by a new bridge at a site 260 yd upstream. The building of the new bridge was commenced in 1789 by William Eboral, the stonemason, and it was completed in 1793 at a cost of £3258. Castle Bridge remains to this day a fine example of bridge building in stone.

The bridge has a single segmental arch of 105 ft span with a rise of 23 ft 9 in. and carries the Warwick to Banbury road (A425) over the Avon. The arch ring is built from massive banded ashlar sandstone blocks and is 3 ft 6 in. deep. The 22 ft wide roadway is flanked by footpaths 6 ft 6 in. wide on both sides and there is an elegant balustraded parapet.

The design of the bridge is clearly based on that by Robert Mylne for an earlier bridge also built for the Warwick estate at Leafield about 1 mile downstream. It is recorded that Mylne presented a report on the proposed Castle Bridge in 1774, although construction did not commence until fifteen years later.

After the completion of Castle Bridge, the Earl incorporated the Great Bridge and its associated roadways into the Castle park. A few arches of the Great Bridge still remain on the north side of the river.

Further reading

Victoria County history of Warwickshire. VIII, OUP, 1969, 463.

23. Oxford Canal

HEW 1612

SP 362 845 to SP 508 064

In 1769 the Act was passed for a new canal from the Coventry Canal at Longford to the River Thames at Oxford, thus completing James Brindley's 'Grand Cross' of

Castle Bridge, Warwick

canals connecting the Severn, Mersey, Trent and Thames rivers across central England. With Brindley as the Engineer, cutting of the canal started at the Coventry end and by August 1774 it was open as far as Napton and by March 1778 to Banbury, 63 miles from Longford. Due to financial problems work stopped at Banbury.

After much argument and discussion between the Oxford, Coventry, Trent and Mersey and Fazeley canal companies, the Oxford company agreed, as part of the 'Coleshill' agreement of 1782, to complete the line of their canal between Banbury and Oxford. Work was re-started in 1786 and the 77 mile long canal was finally opened to traffic on 1 January 1790, 20 years after construction started.

The canal rises 74 ft 4 in. by means of 13 locks from the Coventry Canal to the summit level at Marston Doles and then falls 195 ft by 30 locks to the River Thames at Oxford. Below the summit the canal joins the valley of the River Cherwell which it follows for 33 miles to Oxford. During its descent of the Cherwell the canal crosses the river on the level twice. Three miles north of Oxford is Dukes Cut, a branch canal to the River Thames opened in 1789. At Oxford the canal enters the River Thames by Isis Lock. The original basin and wharves in Oxford were sold in 1937 and filled in.

Oxford Canal

The line of the canal between Longford and the south end of the summit was constructed on a very sinuous alignment, following the contours of the land very closely. This is not typical of Brindley's earlier canals, but he died in 1772 during the construction of the canal and it has been suggested that the sinuous course of the canal is due to Samuel Simcock, Brindley's successor as Engineer. During the period 1829–34 much of the original line was straightened out between Hawkesbury and Braunston, reducing the length of the canal by no less than 13 ½ miles, but a good example of the original alignment can still be seen at Wormleighton (SP 43 55).

Other works associated with the Oxford Canal which are described individually in this volume are: Cosford Aqueduct (HEW 1720) and Braunston Footbridges (HEW 352).

Further reading

Compton H. J. *The Oxford Canal*. David and Charles, Newton Abbot, 1976.

24. Fenny Compton 'Tunnel'

HEW 38

SP 433 525 to 442 520

In order to carry the line of the Oxford Canal through high ground near Fenny Compton a tunnel 9 ft wide and 12 ft high was driven for a distance of 1138 yd. The tunnel was opened in 1776. It was provided with a number of passing places 16 ft wide and rings were mounted in the tunnel walls to assist boatmen in pulling boats through the tunnel.

In 1838 the land over the top of the tunnel was purchased by the Oxford Company with a view to opening out the shallow tunnel. The first stage, carried out between 1838 and 1840 involved the opening out of both ends of the tunnel and a short section in the centre to give two separate tunnels of 336 and 452 yd respectively. Later, in 1865 a decision was taken to open out the rest of the tunnel; the opening out of the southern tunnel was completed in 1868 and the northern tunnel in 1870.

The canal now runs through a deep cutting and there is little evidence of the former tunnel. During the opening out several bridges were constructed including a cast-iron roving bridge carrying the towpath across the canal, a bridge carrying the A423 Southam to Banbury road (recently rebuilt in reinforced concrete) and a rectangular wrought iron trough carrying a stream feeding Wormleighton Reservoir.

Further reading

COMPTON H. J. *The Oxford Canal.* David and Charles, Newton Abbot, 1976, 24–25, 102–103.

Site index

Additional sites

1. Derbyshire and Nottinghamshire

Carsington Reservoir, HEW 1902, SK 244 505

Cat and Fiddle Windmill, Ilkeston, HEW 779, SK 437 398

Dunham Bridge, River Trent, HEW 358, SK 819 745

Dunham Pipe Bridge, HEW 1576, SK 819 744

George Stephenson's Sighting Tower, HEW 67, SK 273 645

Heage Windmill, Ripley, HEW 778, SK 367 507

Leahurst Aqueduct, Cromford Canal, HEW 284, SK 320 556

Longcliffe Bridge, HEW 287, SK 235 592

Papplewick Beam Engines, HEW 368, SK 583 521

2. Humberside and North Lincolnshire

Alford Windmill, HEW 771, TF 457 766

Beverley and Barmston Drain, HEW 934, TA 080 500 to TA 100 296

Beverley Minster, Jacking North Transept, HEW 1190, TA 044 395

Boston Haven Swing Bridge, HEW 1666, TF 327 431

Boston Market Place Boreholes, HEW 1801, TF 327 441

Caistor Canal, HEW 1618, TF 011 990 to TF 071 992

Cleethorpes Pier, HEW 1156, TA 308 089

County Bridge, Brigg, HEW 1697, SE 998 072

East Holmes Bridge, Lincoln, HEW 1660, SK 973 710

Fiskerton Sluice, HEW 1766. TF 089 712

Fledborough Viaduct, HEW 1036, SK 816 716

Grimsby – Cleethorpes Coastal Protection, HEW 1292, TA 274 107 – TA 277 114 – TA 302 097

Horncastle Canal, HEW 1842, TF 194 571 – TF 212 577 – TF 255 694

Keadby Canal Drawbridge, HEW 636, SE 826 115

Kent Bridge, Keadby, HEW 1592, SE 841 106

Kirton Lindsey Tunnel, Gainsborough, HEW 1714, SE 938 001 to SE 947 010

Lincoln West Area Drainage, HEW 1833, SK 914 720 to SK 969 703

Littleborough Ford, River Trent, HEW 1252, SK 824 824

Maud Foster Sluice, Boston, HEW 1779, TF 335 430

New Holland Pier, HEW 1243/1244, TA 080 245

Roman Aqueduct, Lincoln, HEW 1690, SK 988 738 to SK 978 720

Skegness Pier, HEW 1732, TF 575 634

Skidby Windmill, HEW 685, TA 020 333

Steeping River Outfall, near Skegness, HEW 1715, TF 543 596 to TF 527 598

St Johns Hospital Water Tower, Bracebridge Heath, HEW 1661, SK 982 677

St Mary's Conduit, Lincoln, HEW 1805, SK 973 704 to SK 977 713

Tattershall Bridge, River Witham, HEW 1818, TF 197 562

Tholesway Water Mill, near Market Rasen,HEW 1757, TF 166 967

Westgate Water Tower, Lincoln, HEW 1577, SK 975 721

Wrawby Post Mill, Brigg, HEW 642, TA 025 088

Yarborough Bridge, Brigg, HEW 1717, SE 993 070

3. South Lincolnshire and Cambridgeshire

Alconbury Bridge, HEW 60, TL 18 75

Fosdyke Bridge, HEW 1724, TF 319 322

Holme Fen Survey Post, HEW 819, TL 202 839

Grantham Conduit, HEW 1647, SK 913 359

Sneaths Windmill, Lutton Gowts, HEW 1722, TF 435 243

Thorney Tank Yard, HEW 1172, TL 280 044

4. Norfolk and North Suffolk

Dukes Palace Bridge, Norwich, HEW 1027, TG 229 088

East Suffolk Railway, HEW 1684, TM 145 445 to TM 547 928

Fakenham Gasworks, HEW 96, TF 919 293

Midland and Great Northern Joint Railway, HEW 1665, TF 01 18 to TG 52 08

St Olaves Windmill, HEW 1566, TM 457 998

Vauxhall Bridge, Yarmouth, HEW 391, TG 521 080

5. South Suffolk and Essex

Blue Mills Bridge, Witham, HEW 1578. TL 830 131

Bures Bridge, HEW 1643, TL 906 340

Colchester, Clacton and Walton Railway, HEW 1663, TM 003 263 to TM 176 154 and TM 252 215

Gunpowder Works, Waltham Abbey, HEW 1442, TL 37 01

Harwich Railway, HEW 1664, TM 096 323 to TM 260 325

Littlebury Bridge, HEW 1755, TL 519 396

London Road Bridge, Chelmsford, HEW 1363, TL 709 066

Shenfield to Southend Railway, HEW 1686, TQ 613 950 to TQ 881 860

Sparrows Bridge, Gosfield, HEW 1851, TL 790 292

6. Hertfordshire, Bedfordshire and North Buckinghamshire

Coal Duty Marker Posts, HEW 1848, TQ 07, TL 50

Denham Court Lenticular Bridge, HEW 1919, TQ 052 873

Leckhampstead Bridge, HEW 1754, SP 738 355

Palladian Bridge, Stowe, HEW 1920, SP 680 372

Welwyn Tunnels, HEW 1926, TL 249 164 to TL 253 173

7. Leicestershire and Northamptonshire

Abbey Pumping Station, HEW 1746, SK 559 067

Flood Relief Scheme, HEW 1841, SK 570 010 to SK 592 070

New Parks Service Reservoir, HEW 1826, SK 549 049

Thornton Dam, HEW 1871, SK 473 072

8. Warwickshire and North Oxfordshire

Fenny Compton Roving Bridge, Oxford Canal, HEW 1817, SP 433 524

Leafield Bridge, HEW 1088, SP 280 630

Leamington Swimming Pool Roof, HEW 1742, SP 318 655

Stratford-upon-Avon Viaduct, HEW 1830, SP 188 533

Wixford Bridge, HEW 1781, SP 087 546

Index of Engineers

Architects

Contractors

General bibliography

Adamson S. H. *Seaside piers*. Batsford, London, 1977.

Bainbridge C. *Pavilions on the sea*. Robert Hale, London, 1986.

Barbey M. F. Civil engineering heritage: northern England. Thomas Telford, London 1981.

Beckett D. *Stephenson's Britain*. David and Charles, Newton Abbot, 1984.

Biddle G. and Nock O. S. *The railway heritage of Britain*. Michael Joseph, London, 1983.

Binnie G. M. *Early dam builders in Britain*. Thomas Telford, London, 1988.

Blower A. *British railway tunnels*. Ian Allan, London, 1964.

Bonavia M. R. *Historic railway sites in Britain*. Rober Hale, London, 1987.

Boyes J. and Russell R. *The canals of eastern England*. David and Charles, Newton Abbot, 1977.

Brown R. J. *Windmills of England*. Robert Hale, London 1976.

Darby H. C. The draining of the Fens. CUP, 1940.

Faulkner A. H. *The Grand Junction canal*. David and Charles, Newton Abbot, 1972.

Gordon D. I. *A regional history of the railways of Great Britain: the eastern counties*. David and Charles, Newton Abbot, 1977.

Hadfield C. *British canals: an illustrated history*. David and Charles, Newton Abbot, 1979.

Hadfield C. *The canal age*. David and Charles, Newton Abbot, 1981.

Hadfield C. *The canals of the East Midlands*. David and Charles, Newton Abbot, 1981.

Hadfield C. and Skempton A. W. *William Jessop, engineer*. David and Charles, Newton Abbot, 1979.

Hague D. B. and Christie R. *Lighthouses, their architecture, history and archaeology*. Gomer Press, Llandyssul, 1975.

Humber W. *Cast iron and wrought iron bridges and girders for railway structures*. Spon, London, 1857.

Jervoise E. *The ancient bridges of mid and eastern England*. The Architectural Press, London, 1932.

Jones E. *The Penguin guide to the railways of Britain*. Penguin, Harmondsworth, 1981.

Leleux R. A. *A regional history of the railways of Great Britain: the east Midlands*. David and Charles, Newton Abbot, 1976.

Long N. *Lights of East Anglia*. Terence Dalton, Lavenham, 1983.

Marshall J. *A biographical dictionary of railway engineers*. David and Charles, Newton Abbot, 1978.

Marshall J. *The Guiness railway book* . Guiness Publishing Ltd, 1989.

Morgan B. *Railways: civil engineering*. Arrow Books, London, 1973.

Penfold A. (ed.) *Thomas Telford: engineer*. Thomas Telford, London, 1980.

Rolt L. T. C. *George and Robert Stephenson*. Pelican, London, 1978.

Sivewright W. J. *Civil engineering heritage: Wales and western England*. Thomas Telford, London, 1986.

Skempton A. W. *The engineering works of John Grundy, 1719–83*. Lincolnshire Historical and Archaeological Soc., **19**, 1984.

Walker C. *Thomas Brassey: railway builder*. Frederick Muller, London, 1969.

Walters D. *British railway bridges*. Ian Allan, London, 1963.

Whishaw F. *The railways of Great Britain and Ireland*. Simpkin Marshall, London 1840; David and Charles, Newton Abbot, 1969.

Metric equivalents

Imperial measurements have been used in giving dimensions of the works described, as this system was used in the design of the great majority, except for modern structures where the appropriate metric units are given.

The following are metric equivalents of the Imperial units used.

Length	1 inch = 25.4 millimetres
	1 foot = 0.3048 metre
	1 yard = 0.9144 metre
	1 mile = 1.609 kilometres
Area	1 square inch = 645.2 square millimetres
	1 square foot = 0.0929 square metre
	1 acre = 0.4047 hectare
	1 sqare mile = 259 hectares
Volume	1 gallon = 4.546 litres
	1 cubic yard = 0.7646 cubic metre
Mass	1 pound = 0.4536 kilogram
	1 UK ton = 1.016 tonnes
Pressure	1 pound force per square inch = 6.895 kilonewtons per square metre = 0.06895 bar
Power	1 horse power = 0.7457 kilowatt

Subject Index